THE ULTIMATE PREPPER'S SURVIVAL GUIDE

THE COMPLETE DIY GUIDE FOR SELF-SUFFICIENCY IN ENERGY, FOOD SUPPLY, WATER FILTRATION, MEDICINE, STOCKPILING, OFF-GRID LIVING, HOMESTEADING, URBAN SURVIVAL, BUSHCRAFT, & MORE...

TATE GRIFFIN

First Edition.

Printed in the USA

TABLE OF CONTENTS

INTRODUCTION

Hello friend. Welcome to *The Ultimate Prepper's Survival Guide*. This is your all-in-one handbook for navigating the unpredictable twists and turns life might throw your way.

If you're someone who's just starting to dip his or her toes into becoming self-sufficient and prepared for emergencies, but if you also don't know where to start, this book will be just what you're looking for.

Let's face it: the world is changing pretty rapidly. Over the last few years we've seen several natural disasters to economic shifts to unforeseen emergencies like the Covid pandemic. On top of that, you can never truly know when a major disaster might strike your town or your local area.

This is why being prepared and boosting your self-sufficiency isn't just a good idea. It's absolutely essential!

You see, it's not crazy to become a prepper. While there's often a stereotype out there that preppers are always paranoid and almost eagerly waiting for the world to end, the truth is that the vast majority of preppers are just normal people like you. The only real difference between preppers and everyone else is that preppers have taken certain steps and actions in their lives to become more prepared for unexpected emergencies. There's nothing crazy about that.

But where do you start? That's where this guide comes in. Think of this book as your trusty companion. It will be beside you as you build a foundation of self-reliance in various aspects of your life and you can refer back to it at any time.

To make things as easy as possible for you, we've broken down the essentials into manageable sections, with each focused on a key area of preparedness and self-sufficiency. Regardless of whether you're looking to secure your energy supply, grow your own food, have clean water, or stockpile other essentials, we've got you covered.

But we won't stop there.

We'll also dive into off-grid living, homesteading, urban survival, bushcraft, and so much more. This book will be a comprehensive approach because true preparedness itself needs to be comprehensive and touch every part of your life.

Imagine waking up each morning knowing that you have the skills and resources to sustain yourself and your loved ones, regardless of external circumstances. Yes, it might sound daunting at first to learn everything, but we'll break everything down into simple and actionable steps that you can implement at your own pace.

There's a lot to consider, and it's easy to get bogged down by information overload. That's why we've focused on making this guide as approachable and practical as possible.

Examples of topics we'll cover in this guide include:

- How to harness solar, wind, and other renewable energy sources to power your home sustainably. We'll examine everything from setting up solar panels to maintaining backup generators.
- Techniques for growing your own vegetables, preserving food, and even raising livestock.
- How to have access to clean water is non-negotiable. We'll guide you through various filtration and storage solutions so you'll never go thirsty.
- How to handle medical emergencies in the event that you don't have access to a hospital or a medical professional.
- What to stockpile, how much to keep on hand, and the best practices for rotating and maintaining your supplies.
- How you can embrace a lifestyle that reduces your reliance on public utilities.
- Essential survival skills like navigation, shelter-building, and foraging so you can stay alive if you ever find yourself out in the wilderness.

The purpose of this book isn't only to inspire you but also to equip you to take control of your own destiny. Preparedness isn't about fear. It's about confidence and peace of mind.

So, grab a cup of coffee, settle in, and let's embark on this journey together.

Let's get started!

BOOK 1:

ESSENTIAL PREPPING PRINCIPLES

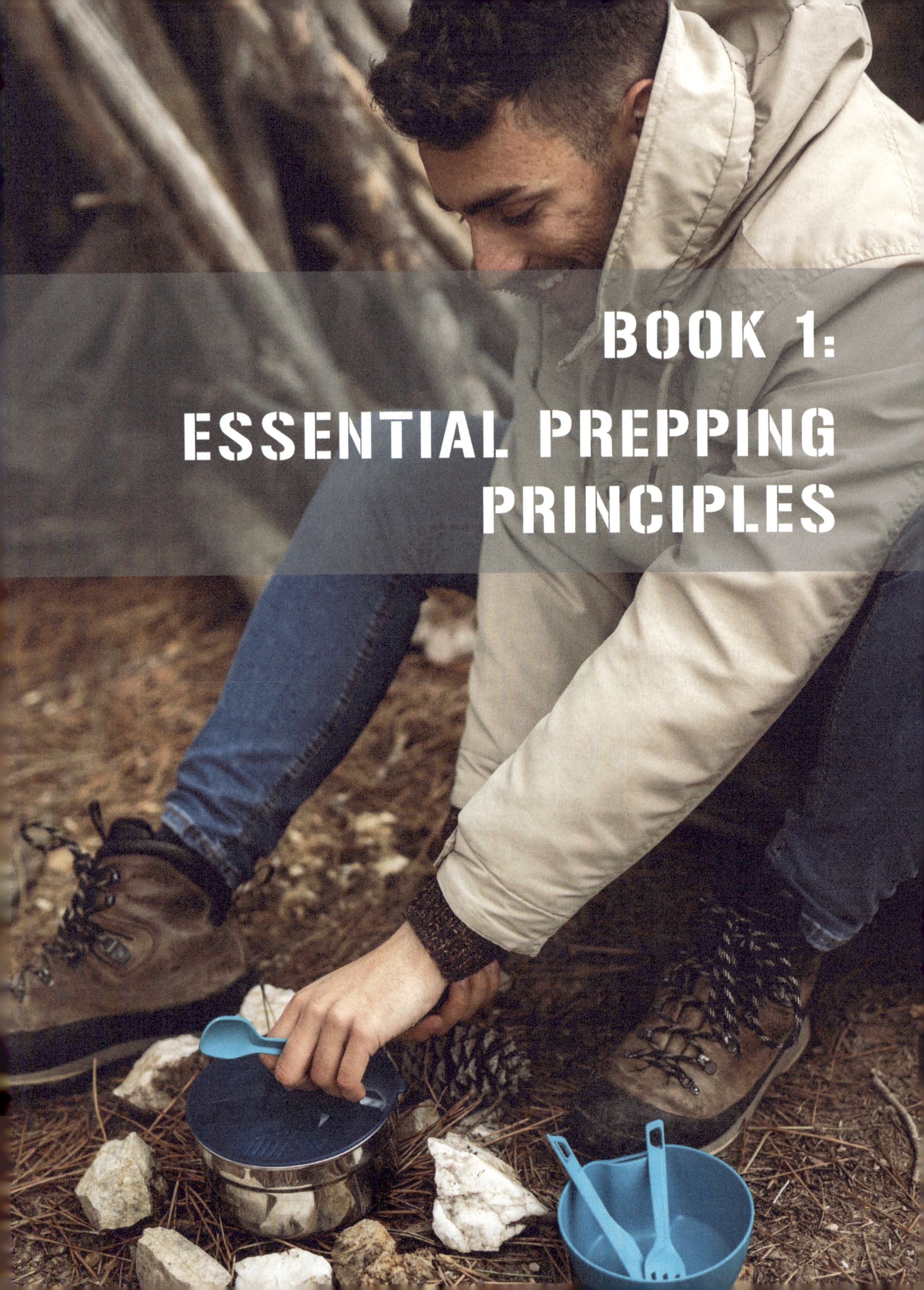

Chapter 1:
Understanding the Necessity of Prepping

Hey there! Welcome to the first book and chapter of "The Ultimate Prepper's Survival Guide."

A few years ago, I would've initially rolled my eyes at the idea of "prepping." It sounded extreme, like something only people on those survival TV shows did. Then Hurricane Harvey hit my area.

Our streets turned into rivers overnight, the grocery store shelves were wiped clean, and the power was out for days. I remember standing in line for hours just to get a case of water and that was when it clicked for me.

This isn't about stockpiling for the apocalypse; it's about being ready for the curveballs life throws at you.

The truth is, disasters are becoming more common and more unpredictable. Whether it's a wildfire spreading faster than expected, a pandemic shutting down entire cities, or just rising prices making everyday basics harder to afford, being prepared isn't overreacting. It's practical. It's about making sure your family has what they need to get through tough times.

And honestly, once you start thinking that way, it feels less overwhelming and more empowering. You're not waiting for help; you're taking control. So yeah, it's important, and trust me, it's easier to start than you think. Let's talk about why.

The Growing Need for Emergency Preparedness

Have you noticed how over the last few years, the news headlines are constantly filled with stories about natural disasters, pandemics, and economic instability?

According to the Federal Emergency Management Agency (FEMA), the number of natural disasters in the U.S. has been on the rise, with events like Hurricane Katrina, wildfires in California, and severe flooding becoming more frequent and intense. Over the past four years alone, FEMA has responded to no fewer than 278 major disaster declarations.

Additionally, global events like the COVID-19 pandemic have shown us how quickly life can change. Furthermore, economic uncertainties (such as inflation and a rising cost of living) have further demonstrated how important it is for ordinary people like you to learn survival skills and have a disaster preparedness plan in place.

With all that being said though, is it really important for you to become a prepper? The short answer is yes, and in the next section, we'll explain why.

Building a Self-Sufficient Mindset

Contrary to what some people think, prepping goes beyond hoarding supplies. It's about fostering a self-sufficient mindset.

You can start by learning basic survival skills such as first aid, water purification, and how to grow your own food. Knowing those skills is just as important as stockpiling medicine, water, and food.

The Centers for Disease Control and Prevention (CDC) recommends everyone have at least a 72-hour emergency kit, but true self-sufficiency involves knowing *how* to use those resources effectively.

In short, a self-sufficient mindset doesn't mean just stockpiling supplies. It also means learning basic survival and preparedness skills.

Chapter 2:
Assessing Risks and Vulnerabilities

Now that we've covered why prepping matters, let's dive into figuring out *what* you need to prepare for.

Identifying Personal and Environmental Risks

First thing's first, not all risks are created equal!

You'll want to start by looking around your environment. Do you live in an area that's prone to hurricanes, earthquakes, or wildfires?

That's just the first question you'll want to ask. Additional questions that you can ask to identify personal and environmental risks include:

- What natural disasters are common in my area?
- What challenges are specific to my location?
- Are there any vulnerabilities in my home or property?
- What are the specific health needs of my household?
- Do I have young children, elderly family members, or pets who need tailored plans or supplies?
- What seasonal risks should I plan for?

Conducting a Family Risk Assessment

Sit down with your family and discuss possible risks and how each person can contribute to the preparedness plan.

When we sat down as a family to go over our risk assessment, it was a little awkward at first, like, where do you even start with something like this? But once we got talking, it actually became pretty productive. For example, my teenage son, who's always glued to his phone, pointed out that he could handle communications during an emergency because he knows how to navigate apps, send alerts, and even find local resources.

My wife, who's way more organized than me, offered to take charge of supplies by keeping an inventory of food, water, and first-aid items. And my youngest, who's only eight, actually lit up when we gave her a role too. She's now our "flashlight manager" and knows exactly where every light source in the house is. By the end of the conversation, it felt less like a "scary prepper thing" and more like a team project where everyone had a part to play.

Honestly, it made us all feel more confident. I'd recommend every family do this, even if it feels weird at first. Basically, just identify roles and responsibilities, as in who will handle first aid, who will manage supplies, and who will communicate with emergency services if needed, among other important tasks.

It's very important to adopt a collaborative approach with your family rather than trying to do everything on your own. Everyone in your family (including your children) should be aware of potential disasters and how each of them can contribute to the family's overarching disaster preparedness strategy. That means each person in your family should be learning survival skills with you.

We'll explore more on this in the next chapter.

Chapter 3:
Developing a Family Survival Strategy

Now it's time to put together a solid family survival strategy.

Creating an Effective Communication Plan

First things first: communication. In any emergency - whether a natural disaster like flooding, an EMP attack, or a solar flare - staying connected with your family is critical.

Here's how to set it up effectively:

1. Choose a family member who lives out of the immediate area as your main contact person. This person will coordinate information if local lines are down.
2. Establish multiple communication methods:
   - Text Messages: These often go through when calls don't. Encourage your family to use texting as a primary method.
   - Walkie-Talkies: Also known as two-way radios, they're ideal for short-range communication if cell networks fail.
   - Messaging Apps: Set up a family group chat on apps like WhatsApp or Telegram for quick updates.
3. Establish a couple of safe meeting locations in case you get separated. These could be a neighbor's house or a local landmark that's easy to find.

Assigning Roles and Responsibilities in an Emergency

Everyone in the family will have unique strengths, so assigning roles can make your response more efficient.

Only you can come up with the specific roles, but here are a few options to give you an idea:

- **The Planner:** This person keeps the emergency kit updated and ensures everyone knows the plan.

- **The Communicator:** This is the individual who oversees all communication tasks, keeps everyone informed, and contacts the designated primary person.
- **The Medic:** This person takes care of first aid needs, knows basic medical procedures, and keeps the first aid kit stocked.
- **The Cook:** The cook manages food supplies, prepares meals during an emergency, and ensures water purification if needed.
- **The Navigator:** This person will be the most familiar with evacuation routes and can lead the family to safety if evacuation is necessary.

The bottom line is this: make sure that each family member has a crystal clear role to play.

Preparing Your Home for Emergency Situations

Your home should be as prepared as possible to handle emergencies. Here are specific steps to take:

1. Anchor heavy furniture to walls to prevent tipping during earthquakes or strong winds that can occur during natural disasters.
2. Make sure you have working smoke detectors and carbon monoxide alarms. Regularly check and replace their batteries.
3. Stock a safe room with essential supplies such as:
    - At least one gallon of water per person per day for three days.
    - Non-perishable items that are easy to prepare.
    - Flashlights and batteries.
    - A comprehensive and easily accessible first aid kit.
    - Blankets and clothing
    - Any necessary prescriptions and a basic medicine supply.
4. Store important documents like passports, insurance papers, and medical records in a waterproof and fireproof container. Keep a digital copy on a secure cloud service for easy access if physical copies are damaged.

There you have it! In short, assign roles to family members, have designated safe areas outside of town and practice your routes for getting there, and have a stockpile of supplies at home in case you can't evacuate.

Prepping is about staying ready for life's surprises, from natural disasters to economic challenges. It's not just about stockpiling. It's also about learning skills like first aid and water purification while involving your family and community. View prepping as making tough times manageable and even improving your day-to-day life.

Key Points To Remember:

1. Disasters are more common, so prepping is practical, not extreme.
2. Focus on the basics first, like food, water, and first aid.
3. Learn skills like first aid, water purification, and gardening. It's not just about supplies.
4. Involve your family in prepping so everyone knows the plan and feels confident.
5. Build community with friends and neighbors for shared knowledge and resources.
6. At minimum, have a 72-hour emergency kit for each family member.
7. Keep prepping simple and adaptable. Start small and build over time.
8. A self-sufficient mindset will get you further than just hoarding supplies.
9. Prepping has everyday benefits, like saving money and staying organized.
10. Take action now, even if it's just one small step. It all adds up!

Speaking of stockpiling supplies, in the next chapter, we'll dive into one of the most important kinds of supplies that you need to stockpile: water.

BOOK 2: WATER SOURCING AND FILTRATION

Chapter 1:
Locating Water Sources

It goes without saying how vital water is to survival. You can only survive three days without it, and it will take less than one day before you begin feeling the negative effects of dehydration.

Finding Urban Water Sources During Crises

Living in a city doesn't mean you're out of options when it comes to finding water during a crisis. If you're at a loss at where you can find water, try these areas:

1. Even if the main supply is disrupted, public buildings like libraries, schools, and community centers often have backup water supplies.
2. In some cases, fire hydrants can be a source of water, but use this as a last resort.
3. If you have access to rooftops, setting up a rainwater collection system with clean containers can provide an additional water source.

During a power outage in my city last summer, we had a brief moment where we realized we weren't sure if water was going to keep flowing. It wasn't full-on panic, but we definitely felt the pressure to figure out what our options were. We ended up hitting a local school because we knew it had backup water tanks for emergencies. The custodian there, who was super kind, told us they were distributing water to families in the neighborhood. Another neighbor mentioned that fire hydrants could technically be used, but they'd need special tools, so we mentally filed that away as a last-resort idea.

What really saved the day, though, was a bit of quick thinking. We had clean buckets and containers, so we set them up on our balcony to collect rainwater since a storm was rolling in. It wasn't much at first, but seeing those containers fill up was a huge relief. That whole experience taught me that even in the middle of the city, you can find water. You just have to think outside the box and be resourceful!

Locating Natural Water Sources in the Wilderness

When you're out in nature, finding water can be more straightforward, but it still requires some knowledge and preparation:

1. Flowing water is generally safer than stagnant sources. Look for signs of flow, like moving water or wildlife activity, which indicate a fresh source.
2. Lakes and ponds are also reliable water sources, but you'll need to filter and purify the water to remove contaminants.
3. Collecting rainwater using tarps or containers is a great option. Dew can also be gathered using absorbent materials like a bandana in the early morning.
4. Research natural springs in your area as well, as they'll also be a great source of water.

Understanding Water Contamination Risks

Not all water is safe to drink, and contamination can come from various sources:

1. Bacteria, viruses, and parasites can make water unsafe. Always purify water from natural sources before drinking.
2. Runoff from industrial areas or agricultural sites can introduce harmful chemicals. Avoid water near these sources if possible.

3. Sediments and debris can clog filters and make purification more difficult. This is why you should utilize pre-filtration methods like cloth straining.

It's also worth noting that water from a flood isn't safe to drink. Assume it's contaminated as well and don't drink it!

Creating a Water Source Map for Emergencies

Having a detailed water source map can save you valuable time during a crisis. Here's how to create one:

1. Mark public water sources, natural water bodies, and potential collection points in both urban and rural areas.

2. Provide clear directions and distances from your home to each water source.

3. Indicate which sources are easily accessible and which might require some effort to reach.

There you go! At the end of the day, you just want to be acutely aware of the best places inside and outside of your local town where you can find water.

The first time I realized how critical it is to know where your water sources are, I was stuck in a ridiculous situation. It was a scorching summer day, and I was in the middle of what I thought was a "routine" off-grid camping trip with a buddy. Now, my buddy and I had been prepping for months by having the right gear, building fire skills, even practicing some rudimentary water filtration techniques. We were feeling pretty smug about our ability to "survive." We knew our food sources, our shelter options, and our route.

But we didn't think enough about *water*. We didn't map out every creek, stream, or well in the area. I mean, how hard could it be to find water in the wilderness, right?

So there we were, about 12 hours into our trip, and we realized we had absolutely *zero* water left. And by zero, I mean *not a single drop*. I started feeling the panic rising as the temperature soared above 90 F.

My buddy, being the eternal optimist, kept telling me, "We'll find something. We're tough."

Well, turns out, we were also really bad at reading the landscape for natural water sources. We wandered for hours, dehydrated and getting cranky, before I finally remembered the importance of having a map for these kinds of situations. At that moment, I thought about how much time (and panic) we could've avoided if I'd just taken the time to mark every accessible water source on a map before heading out.

So, take it from me: don't skip this step. And while you're at it, consider this a reminder that water isn't just something you find by chance. It's something you plan for.

Chapter 2:
Water Filtration and Purification Techniques

Knowing how to find water is super important, but at the same time, drinking water that's been contaminated can sometimes be more dangerous than drinking no water at all. Fortunately, there are many methods at your disposal to make sure that the water you collect will be safe to drink.

Boiling and Chemical Treatment for Purification

Boiling is one of the simplest and most reliable methods to purify water. Here's how to do it effectively:

1. Boiling:
    - Ensure the water reaches a rolling boil for at least one minute (three minutes at higher altitudes).
    - Let the water cool naturally and store it in clean, covered containers.
    - Always use a pot with a lid to speed up the boiling process and conserve fuel.
2. Bleach
    - Regular household bleach (unscented) can disinfect water. Add 8 drops per gallon, stir, and let it sit for 30 minutes. If the water is cloudy, double the amount of bleach.
3. Water Purification Tablets
    - These are convenient and portable. Most purification tablets require about 30 minutes to take effect.

UV Water Purification and Portable Filters

Modern technology offers some fantastic water purification options that are both effective and easy to use:

1. UV Purifiers
   - UV light disrupts the DNA of bacteria, viruses, and other pathogens, which makes them harmless. Simply submerge the UV device in the water for the recommended time (usually around 90 seconds).
2. Portable Filters
   - Compact and lightweight, straw filters like the LifeStraw allow you to drink directly from a water source. These systems use gravity to push water through a filter, requiring minimal effort.

DIY Filtration Methods: Sand, Charcoal, and Cloth

If you're caught without commercial filters, don't worry. You can still make your own water filtration system using common materials:

1. Sand and Gravel Filter:
   - Gather sand, gravel, and a container with small holes.
   - Then follow these steps:

1. Layer gravel at the bottom of the container.
2. Add a thick layer of sand on top.
3. Pour water through the filter, allowing the sand and gravel to trap larger particles.
4. Boil or chemically treat the filtered water to ensure safety.

2. Charcoal Filter:
   - You'll need activated charcoal (available in aquarium stores), sand, gravel, and a container.
   - Then complete these steps:
     1. Layer gravel, then sand, followed by activated charcoal in the container.
     2. Pour water through the layers to remove impurities and improve the quality of the taste.
     3. Boil or chemically treat the water after filtration.
3. Cloth Filtration:
   - All you'll need is a clean cloth or bandana.
   - Then complete these steps:
     1. Fold the cloth several times to create multiple layers.
     2. Pour water through the cloth to remove larger particles.
     3. Follow up with boiling or chemical treatment for purification.

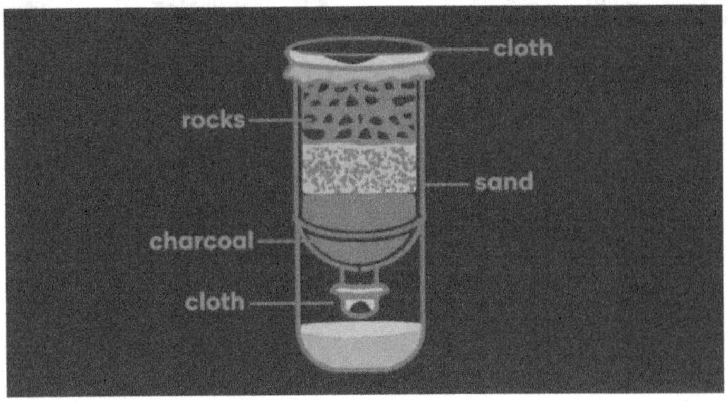

Source

Choosing the Best Commercial Filtration Systems

When it comes to commercial filtration systems, there are plenty of options, but each are tailored to different needs. Here's what to think about:

1. Filter Type
    - **Activated Carbon Filters:** These are excellent for improving taste and removing chemicals but might not eliminate all pathogens.
    - **Ceramic Filters:** These are effective against bacteria and protozoa but may require regular cleaning.
    - **Reverse Osmosis Systems:** These kinds of filters are highly effective but typically require a power source and are less portable.
2. Portability
    - **Personal Filters:** These filters are compact and lightweight, ideal for individual use during hikes or short trips. You can easily stash them in your backpack, for instance.
    - **Family-Size Filters:** These are larger systems designed to provide water for multiple people; you can keep them in your kitchen.
3. Maintenance and Longevity
    - Check how long the filter lasts before needing replacement. It's also your responsibility to research the maintenance tips for any specific types you choose to use.

There are no shortage of options when it comes to water filters, as you can probably tell. As a golden rule, you'll want to have at least one family-sized water filter and a personal water filter for each member of your family.

Chapter 3:
Storing Water for Survival

Alright, let's talk about one of the most crucial aspects of prepping: storing water. You've got your sources lined up and your purification methods ready, but without proper storage, all that effort might go to waste.

Best Practices for Water Storage

First things first, let's cover the basics of storing water effectively:

1. Always use clean, food-grade containers for storing water. Avoid containers that have previously held chemicals or non-food items; residues can contaminate your water supply.
2. Heat and sunlight can degrade water quality over time. That's why you should choose a storage spot that stays cool and out of direct sunlight, like a basement or a cupboard.
3. Label each container with the date of storage and the contents to keep track of freshness.
4. Even with proper storage, water doesn't last forever. Aim to rotate your water supply every six months to ensure you always have fresh water on hand.

How to Store Water Long-Term: Containers and Systems

When it comes to long-term water storage, choosing the right containers and systems can make all the difference. Here's what you need to know:

- Food-Grade Plastic Containers
  - These are a popular choice for water storage. Look for containers labeled as food-grade (typically #2 or #4 plastics) to ensure they're safe for long-term use.
- Glass Jugs

    - Glass is non-reactive and won't leach chemicals into your water. However, they're heavier and more fragile, so they're best for smaller quantities or indoor storage.

- Stainless Steel Containers
    - These are durable and long-lasting, which makes them an excellent option for both short-term and long-term storage. They're also less likely to develop mold or algae.
- Water Barrels
    - For those of you who are looking to store larger quantities, water barrels are a great option. They can hold 55 gallons or more and are perfect for families or groups.

Ensuring Water Purity Over Time

Here are some tips to keep your water clean:

1. Adding a small amount of unscented household bleach (about 1/8 teaspoon per gallon) can help prevent the growth of bacteria and algae in your stored water.
2. Always keep your water containers tightly sealed to prevent contamination from dust, insects, or other pollutants.
3. Regularly inspect your water storage for any signs of contamination, such as cloudiness, unusual smells, or floating debris. If you notice anything off, purify the water before use.
4. Keep the area where you store your water clean and free from pests. To do this, just regularly wipe down surfaces and ensure that containers are stored off the ground to prevent moisture buildup.

Project: Build a Solar-Powered Water Distiller

Now, let's get hands-on with a cool project that can enhance your water storage and purification setup: a solar-powered water distiller.

Here's what you'll need:

- A large glass or metal bowl
- A smaller glass or metal container (to collect distilled water)
- Clear plastic wrap
- A small rock or weight

- Tubing (optional, for directing water flow)

Now follow these steps:

1. Place the large bowl in a sunny spot. Pour your contaminated water into the bowl, making sure it doesn't overflow into the smaller container.
2. Place the smaller container in the center of the large bowl. It should sit above the water level so that only the vapor can condense into it.
3. Stretch the clear plastic wrap tightly over the top of the large bowl. Make sure it's sealed all around to prevent vapor from escaping.
4. Place a small rock or weight in the center of the plastic wrap directly above the smaller container. This will create a low point where the distilled water can drip into the collection container.
5. As the sun heats the water, it will evaporate, condense on the plastic wrap, and drip into the smaller container as purified water.

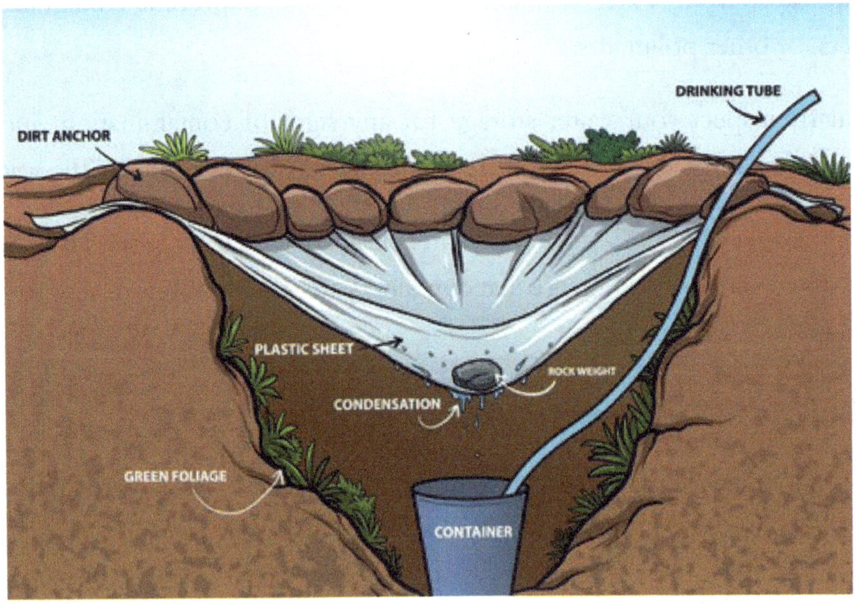

Source

It really doesn't need to be said that water is absolutely essential for survival, and knowing where to find it (whether you're in the city or out in the wild) can make all the difference. In urban areas, places like public buildings or rainwater collection systems can save you in a pinch. In the wilderness, look for flowing water or natural springs, but always be aware of contamination risks.

Having a plan and a water source map ready will make things a whole lot easier in an emergency.

Key Points to Remember:

1. Public buildings like schools or community centers often have backup water supplies you can access during a crisis.
2. Fire hydrants are another option, but they should be a last resort since they might require special tools.
3. If you're in a city, setting up a rainwater collection system on a rooftop with clean containers is a great way to gather water.
4. In the wilderness, flowing water like streams or rivers is usually safer than stagnant ponds, and looking for wildlife activity can help you find fresh sources.
5. You can also collect rainwater with a tarp or gather morning dew with a bandana.
6. Whatever the source, always filter and purify the water before drinking to avoid bacteria or chemicals.
7. Steer clear of water near industrial areas or farmland runoff, as it's likely contaminated.
8. Make a water source map ahead of time so you know exactly where to go if the need arises.
9. Rotate stored water every few months and keep an eye out for contamination so you're always ready.

In the next chapter, we'll dive into something that's just as important as water: storing and preserving food over the long-term.

BOOK 3: LONG-TERM FOOD STORAGE AND PRESERVATION

Chapter 1:
Canning and Food Storage

Canning is one of the best ways to preserve food for the long haul. In this chapter, you'll learn how to do it:

Introduction to Canning and Its Benefits

Canning works by sealing food in airtight jars and heating them to kill off bacteria and other spoilage agents. This process locks in freshness and flavor while extending the shelf life of your food by months….or even years.

It's basically an excellent way to use up surplus produce from your garden or snagged deals from the farmer's market.

Water-Bath Canning vs. Pressure Canning

There are two types of canning, and knowing which one to use is critical for safety.

- **Water-Bath Canning** is best for high-acid foods like fruits, jams, jellies, pickles, and tomatoes. These foods are naturally acidic enough to inhibit the growth of dangerous bacteria like botulism. The process is simple: pack your food into jars, cover them with boiling water in a canner, and boil for the time specified in your recipe.

- **Pressure Canning** is required for low-acid foods like vegetables, meats, poultry, and soups. These foods need higher temperatures to kill off bacteria, which is why a pressure canner is essential. The steam inside the sealed canner needs to reach up to 240°F.

Essential Equipment for Safe Canning

To get started, you'll need a few key items:

- Use glass Mason jars with two-piece lids.

- New lids for each can to ensure a proper seal. Bands can be reused if they're rust-free and in good condition.
- A water-bath canner for high-acid foods and a pressure canner for low-acid foods.
- A jar lifter
- A funnel
- A simple tool (or a clean spatula) to eliminate air bubbles, ensuring an airtight seal.

Preparing and Canning High-Acid and Low-Acid Foods

High-acid foods are the easiest place to start. Wash and prepare your ingredients before you pack them into jars, leaving about 1/4 inch of headspace. Then you can pour in hot liquid (like syrup, juice, or water), remove air bubbles, and wipe the rim of the jar clean before sealing with a lid and band.

Low-acid foods like green beans or chicken require a bit more attention. First, cook or blanch the food as needed. Then pack it into jars, and leave an inch of headspace. Fill with hot broth or water, remove bubbles, and seal the jar.

Chapter 2:
Dehydration and Freeze-Drying

Dehydration and freeze-drying are fantastic methods for long-term food storage. They both reduce moisture content, which slows spoilage and makes food lightweight and easy to store.

The Benefits of Dehydrating Foods

Dehydration is one of the oldest and simplest ways to preserve food. By removing water, you prevent bacteria, mold, and yeast from thriving. Here's why it's worth considering:

- Properly dehydrated food can last for months or even years when stored correctly.
- Dehydrated food is perfect for bug-out bags, camping trips, or just saving pantry space.
- Dehydration retains most of the food's nutrients while intensifying flavors. Think sweet dried fruits or savory jerky!

Freeze-Drying vs. Dehydrating: What's the Best Option?

Both methods are great, but they have some key differences:

- Dehydration removes 80-90% of moisture by applying heat. The food becomes smaller and chewy or crisp, depending on the item. It's great for fruits, vegetables, and meats like jerky.
- Freeze-Drying removes up to 98-99% of moisture by freezing the food and then using a vacuum to extract water. It retains the food's original texture, flavor, and nutritional value better than dehydration, which makes it ideal for preserving full meals, dairy, and delicate fruits like berries.

I'll admit it: when I first started prepping, I wasn't exactly a freeze-drying aficionado. I thought dehydration was the way to go. It was easier, faster, and, to be honest, it felt a little more like "survivalist" than freeze-drying, which always seemed a bit like some kind of fancy science experiment. I was out in the field, armed with my dehydrator, making fruit leather and beef jerky like a pro, thinking I had it all figured out.

Unbeknownst to me, I was missing out on a powerful food preservation technique.

The turning point came one hot summer when I decided to make a batch of homemade backpacking meals. I had dehydrated the veggies, the meat, and even the rice, but when I went to rehydrate the meal in the middle of nowhere, things didn't go exactly as planned. The veggies were chewier than I would've liked, and the meat? Well, let's just say it wasn't winning any flavor awards. It tasted... like, well, dehydrated meat. Not my proudest moment.

That's when I really started thinking about the whole freeze-drying process, which, at that time, seemed like some futuristic method only for super preppers with deep pockets.

After doing a little research and talking to some friends who had made the jump to freeze-drying, I was sold. The food retains *so much more* of its flavor and texture. I'm talking about full meals that actually taste like the meal you prepared, rather than some desiccated version of it. Freeze-drying preserves the food's nutrients, too, which means you're not sacrificing all the good stuff in the name of preservation.

How to Dehydrate Fruits, Vegetables, and Meats

Dehydration is easy to do at home with a dehydrator or even a standard oven. Here's how to get started:

- Fruits
    - Wash, peel (if necessary), and slice evenly. Apples, bananas, and mangoes are great choices. You can soak fruits like apples in a lemon-water solution to prevent browning. Then arrange the slices on a dehydrator tray and dry at 125–135°F for 8–12 hours until leathery.
- Vegetables
    - Blanch vegetables like carrots, green beans, and broccoli for a few minutes to preserve their color and more importantly their nutrients. Then you can spread them evenly on trays and dry at 125°F for 8–10 hours until brittle.
- Meats (Jerky)
    - Use lean cuts of meat to avoid spoilage. Slice thinly, marinate if desired, and dry at 145–155°F until firm and slightly chewy (usually 4–6 hours). Always ensure meat is fully cooked or heat-treated before dehydration to kill bacteria.

Project: Build a DIY Freeze-Drying System

While commercial freeze dryers are expensive, you can create a simple DIY setup for smaller-scale use.

- Here's what you'll need:
    - A chest freezer or deep freezer
    - Vacuum pump (or a high-powered vacuum-sealing device)
    - Airtight freezer-safe containers or bags
- Then you can follow these steps:
    - Slice the fruits or vegetables thinly and arrange them in a single layer on trays.
    - Place the trays in your freezer. Leave them for 1–2 days until frozen solid.

- Using the vacuum pump or sealer, create a vacuum environment around the food to slowly extract moisture while it remains frozen. This process mimics the freeze-drying cycle.
- Once it's dry, you can transfer the food to vacuum-sealed bags or airtight containers. Label with the date for easy rotation.

Chapter 3: Alternative Preservation Methods

Let's talk about a few less traditional, but totally awesome, ways to preserve food.

The Science Behind Fermentation

Fermentation is like magic. It uses good bacteria to turn sugars into lactic acid, which preserves food naturally. It's what gives sauerkraut, kimchi, and pickles their tangy flavor, and as a bonus, these foods are loaded with probiotics. Gut health, anyone?

Here's how to get started:

- Pick your veggies. Cabbage, carrots, cucumbers…whatever you like.
- Mix 1–2 tablespoons of salt per quart of water.
- Stuff your veggies into a clean jar and pour the brine over them. Make sure everything is completely submerged.
- Use a lid or cloth that lets gases escape but keeps bugs and dust out.
- Let it sit at room temperature for a few days to weeks. Once it's tangy enough for you, pop it in the fridge to slow down the process.

Fermented foods can last for months in the fridge, and they're crazy flavorful. Trust us, once you try it, you'll want a whole shelf of this stuff!

Vacuum Sealing for Long-Term Food Storage

Vacuum sealing is like giving your food a little space suit. It sucks out all the oxygen, so mold and bacteria have a harder time ruining your stash. It's perfect for dry goods, fresh produce, and even frozen meats.

Here's the game plan:

- Use a decent vacuum sealer, which doesn't cost much and will save you a ton of hassle later.
- Now prep your food:
  - Portion out dry goods like rice and beans.
  - Wash and slice veggies for easy use later.
  - Pre-freeze meats to avoid sealing up any liquid messes.
- Seal it up. Follow the instructions on your sealer. Get all the air out, seal it tight, and you're good to go.

From there, keep vacuum-sealed items in a cool, dark, and dry place for maximum shelf life. Adding oxygen absorbers can give you extra peace of mind for long-term storage.

Storing Foods in Root Cellars

Root cellaring is one of the oldest tricks in the book, and it still works like a charm. It's perfect for storing fresh produce like potatoes, carrots, and apples for months without electricity.

Here's what to do:

- If you don't have a traditional root cellar, a cool corner of your basement or even an underground pit can work.
- Aim for 32–40°F and 85–95% humidity. Use a thermometer and hygrometer to keep an eye on things.
- Pack root veggies like carrots and potatoes in bins with sand or sawdust to keep them moist. Just keep apples away from other produce…they release ethylene gas that speeds up spoilage.
- Make sure there's airflow to prevent mold and rot.

Project: Build a Fermentation System for Home Use

If you're ready to up your fermentation game, building a DIY fermentation system is easier than you think. Here's how to do it:

- Get your materials
    - You'll need a large food-grade bucket or crock, an airlock lid (you can find these online), and some weights.
- Chop or shred your veggies, add salt, and mix it all up to start drawing out the juices.
- Stuff the veggies into the crock or bucket and press them down. Make sure everything is submerged in its juices or brine.
- Pop the airlock lid on to let gases escape without letting anything nasty in.
- Keep the crock in a cool, dark spot and check it every few days. Once the fermentation is done, you can then transfer your food to smaller jars and refrigerate.

When I first tried fermentation, it was honestly a bit of a mess. I decided to make sauerkraut because, you know, how hard could it be? I shredded a huge head of cabbage, salted it like the YouTube video said, and packed it into a jar.

But here's the thing, I didn't fully submerge the cabbage in the brine. A couple of days later, I had this weird layer of mold on top, and I thought I'd ruined it.

Turns out, I could just scrape that off (thank you, internet!), and the rest was totally fine.

The end result? The tangiest, crunchiest kraut I'd ever had, and I felt like a fermentation pro. Now I've got jars of the stuff lined up in my fridge. It's weirdly addictive!

In short, by sealing food in airtight jars and using heat to kill bacteria, you can keep surplus garden produce or market deals fresh for months or even years.

Key Points to Remember:

1. Canning preserves food by creating an airtight seal and using heat to kill bacteria.
2. Use water-bath canning for high-acid foods like fruits, jams, and pickles.
3. Pressure canning is essential for low-acid foods like vegetables, meats, and soups.
4. Glass Mason jars, fresh lids, a canner, and basic tools like a jar lifter and funnel are must-haves for safe canning.
5. Fermentation is a simple and natural way to preserve foods like sauerkraut and kimchi while boosting gut health.
6. Vacuum sealing keeps food fresh longer by removing oxygen, making it perfect for dry goods and prepped meats.
7. Root cellars or cool storage spaces are ideal for keeping produce like potatoes and carrots fresh for months.
8. Rotate stored food regularly and always label jars with dates to ensure freshness.
9. Practice and preparation help you perfect these techniques—start small and build confidence.
10. These methods don't just store food; they also help reduce waste and save money in the long run.

There's more to know about food and water and we'll dive into more specific information on gathering both later in this book. Next, however, we'll discuss another very big component of self-sufficiency: generating your own power.

BOOK 4: SOLAR AND OFF-GRID ENERGY SYSTEMS

Chapter 1:
Harnessing Solar Power

Solar power is, without question, the biggest game changer when it comes to off-grid energy. It's renewable and can keep your devices running and lights on no matter what's going on in the world.

Many people don't realize just how easy it is for the power to go out. A simple hurricane, flash flood, or wildfire (even if it's on a small scale) is all it takes to disrupt and shut off the power over a very wide area, at which point, you'll be reliant upon yourself to generate your own power.

Understanding Solar Energy Basics

Here's the deal: solar panels soak up sunlight and turn it into electricity. Those panels are made of photovoltaic (PV) cells that produce direct current (DC) power. But your appliances and devices use alternating current (AC), so you'll need an inverter to convert it. And to keep things running when the sun's not out, you'll also need batteries to store the extra energy.

Here are the basic terms you need to know:

- Watts (W)
  - This measures how much power something can produce or use.
- Kilowatt-hours (kWh)
  - This measures how much energy you're using over time (like on your electric bill).
- Amp-hours (Ah)
  - This indicates how much energy your battery can store.

Choosing the Right Solar Panels and Batteries

Your panels and batteries are the core of your setup.

For solar panels, here's what you're looking at:

- Monocrystalline Panels
    - The most efficient and long-lasting. They're perfect if you don't have much space but are okay with spending a bit more.
- Polycrystalline Panels
    - Slightly less efficient but more affordable. Great if you've got room to spare.
- Flexible Panels
    - Lightweight and portable. Ideal for RVs or temporary setups.

For batteries, you have other options as well:

- Lead-Acid Batteries
    - The budget option. They're reliable but heavy and need regular maintenance (like topping off water).
- Lithium-Ion Batteries
    - These are pricier upfront but last longer, need less maintenance, and take up less space.
- Nickel-Iron Batteries
    - Built like tanks, nickel-iron batteries are super durable and long-lasting. They're a bit bulky and expensive, though.

If you're planning a serious off-grid system, lithium-ion is usually your best bet for balance between cost, efficiency, and ease of use.

Designing and Installing a Solar Power System

Setting up your solar power system isn't as complicated as it sounds, and you shouldn't need to hire an expert to do it.

Here's what you'll need to do:

1. Add up the wattage of everything you'll use daily. Then, multiply by how many hours each thing runs to get your total daily energy use in watt-hours (Wh).
2. Divide your total energy use by the average number of sun hours your area gets in a day. That tells you how much solar panel wattage you'll need.
3. Multiply your daily energy use by the number of backup days you want, and that's your battery storage capacity in watt-hours or amp-hours.
4. Make sure it can handle the total wattage of everything you'll be powering at the same time.
5. Place your panels in the sunniest spot you've got, usually a south-facing roof or a ground-mounted system should be best. Secure everything properly with mounting brackets, and double-check all wiring.

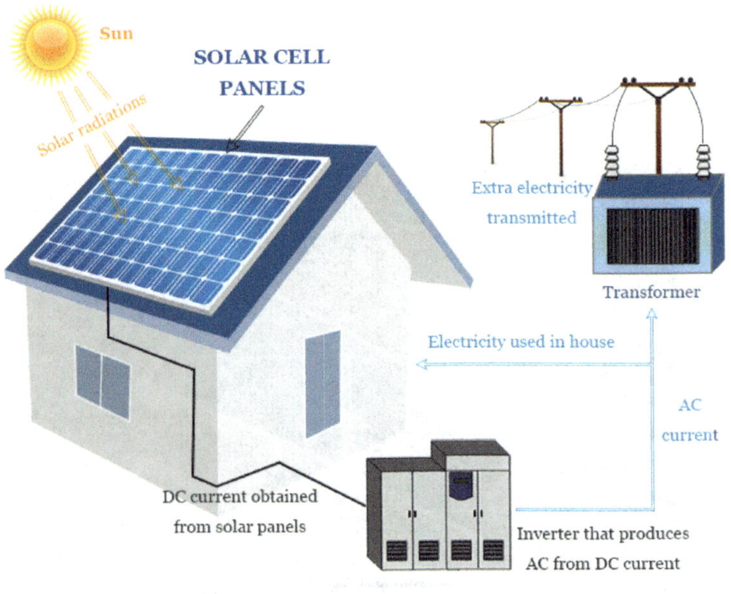

Source

My friend Mark built his own solar setup after a week-long power outage left his family in the dark, and it's proof that anyone can do it. He started small with two monocrystalline panels, a lithium-ion battery, and a basic inverter.

After calculating their daily energy needs (just lights, charging devices, and a mini-fridge) he installed the panels on his shed roof and connected everything. It worked like a charm, and it even powered a portable panel for RV trips. Now, his setup runs their essentials and keeps them ready for anything, and he swears it's easier than it looks.

Chapter 2:
Wind and Micro-Hydro Power

When it comes to off-grid energy, solar power might get all the attention, but wind and micro-hydro power are just as powerful...literally.

Wind energy is a perfect partner for solar power. Why? Because the wind often blows when the sun isn't shining, like at night or during cloudy weather. This means wind turbines can provide consistent energy to complement your solar system. Plus, unlike solar power, it's not dependent on the daylight hours to work.

That said, wind energy works best in areas with average wind speeds of 10–15 mph. If your area is calm most of the time, this may not be the best option.

Building and Installing Small-Scale Wind Turbines

Setting up a wind turbine might sound complicated, but small-scale turbines designed for home use are relatively straightforward.

1. Look for turbines rated to produce enough power for your needs. A small 1-kilowatt turbine is great for basic setups, while a 10-kilowatt turbine can power an entire home.

2. Wind turbines need to be installed in open areas free from obstructions like trees or buildings. Taller is better. As a golden rule, aim for at least 30 feet above anything within a 300-foot radius.

3. You'll need a sturdy pole or tower to hold the turbine. Make sure it's anchored deeply into the ground or bolted to a solid structure.

4. Wind turbines produce DC electricity, so you'll need an inverter to convert it to AC for your appliances. Then you can combine it with batteries to store excess energy.

Micro-Hydro Power: Generating Electricity from Running Water

If you have access to a stream or river on your property, micro-hydro power is one of the most efficient and reliable ways to generate electricity. Unlike solar or wind, it can run 24/7.

Here's how it works:

- A small intake system diverts water from the stream into a pipeline.
- The water flows through a turbine, which spins it to generate electricity.
- The electricity flows into batteries or directly powers your devices through an inverter.

Take note that even a small stream with a few feet of drop can produce usable energy. You don't need a whole river for this method to work.

Project: Build a Micro-Hydro Power System for Your Home

If you've got a stream or river nearby, here's how you can build a simple micro-hydro power system:

1. Measure the flow rate (how much water passes by per second) and the head (the vertical distance the water drops). The more flow and head, the more power you can generate.

2. Gather your materials. You'll need:
   - A turbine (Pelton wheels are a popular choice).
   - PVC or steel pipe for the intake and penstock (the pipeline that channels water to the turbine).
   - An inverter to convert DC to AC power.
   - Batteries for storage.
3. Install a small dam or intake box to direct water into the penstock while keeping debris out.
4. Place the turbine at the bottom of the drop where the water exits the penstock. Then connect the turbine to your power system.
5. Use waterproof cables to connect the turbine to your inverter and batteries.

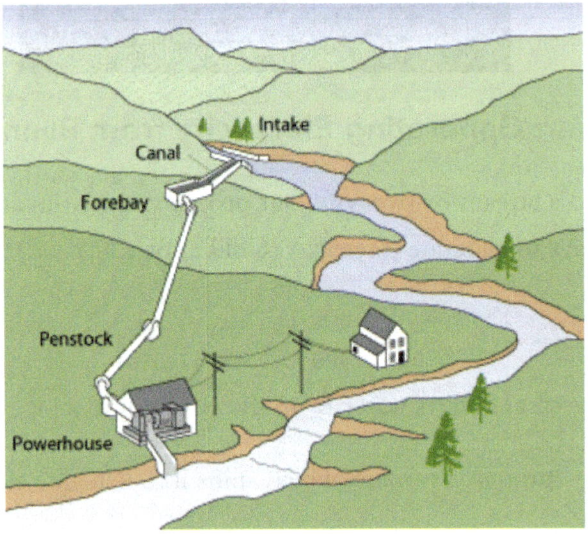

Source

Chapter 3:
Biomass and Alternative Energy Solutions

When it comes to off-grid living, thinking beyond the usual solar and wind power options can open up some exciting possibilities. Biomass and other alternative energy sources like geothermal and tidal power offer unique ways to generate energy and heat.

What is Biomass and How Can It Be Used?

Biomass is any organic material (like wood, agricultural waste, or even animal manure) that can be burned or converted into energy. It's been used for centuries as a heat and energy source.

Here are some common uses:

- Burn wood, crop residues, or compressed biomass pellets in stoves and heaters to warm your home.
- Cooking. Biomass-powered rocket stoves are great for both off-grid living and camping.
- Advanced systems can convert biomass into biofuels, which are then used to generate electricity.

Biomass is especially handy if you're in a rural area or homestead where wood or other organic waste is plentiful.

Constructing Efficient Rocket Stoves and Biomass Heaters

Rocket stoves are a super-efficient way to turn biomass into heat for cooking or warming small spaces. They use a simple design that maximizes combustion, so you burn less fuel while getting more heat.

You'll need fireproof bricks, some metal piping, and insulation material (sand or ash).

Then you can follow these steps:

1. Use the bricks to build a U-shaped burn chamber where the biomass fuel (sticks, pellets) will combust.
2. Attach a vertical pipe or tube at the top of the burn chamber to direct the heat and flames upward.
3. Light a small fire, feed in biomass, and watch how efficiently it burns.

Project: Build a Wind-Powered Generator

If you're up for a hands-on challenge, building a DIY wind-powered generator is a fantastic way to generate off-grid electricity.

First, gather the following materials:

- A small DC motor
- PVC pipes (to create blades)
- A wooden or metal stand (for the base)
- A rectifier (to convert AC to DC if your motor produces AC)
- A battery or storage system

Then you can follow these steps:

2. Cut the PVC pipe into three equal sections, about 2–3 feet long, and shape them into curved blades (cutting with a handsaw is the best way to go about this). Then attach them to the motor's rotor using bolts or a hub.
3. Mount the motor onto a tall, stable base or pole to get it into the wind stream.

4. Attach the motor's wires to the rectifier to convert the energy into a usable form. From the rectifier, you can run the wires to your battery or directly to an inverter.

5. Place the turbine in an open area with consistent wind.

6. Add a tail to the turbine to keep it oriented toward the wind, and install a charge controller to protect your battery from overcharging.

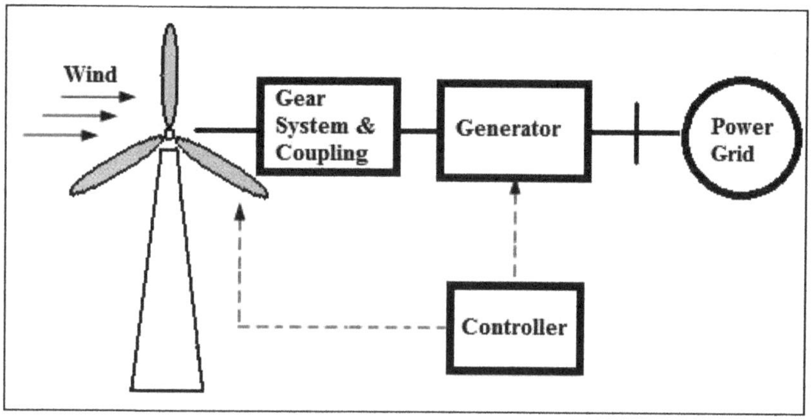

Source

Key Points to Remember:

1. Solar panels convert sunlight into direct current (DC), which is then converted to alternating current (AC) using an inverter.

2. Monocrystalline panels are the most efficient, while polycrystalline panels are affordable, and flexible panels are portable for temporary setups.

3. Lithium-ion batteries are the best choice for off-grid systems due to their efficiency, longevity, and low maintenance.

4. Calculate your daily energy needs and match your solar panel wattage and battery capacity accordingly.

5. Install panels in the sunniest location, like a south-facing roof, and ensure they're securely mounted.

6. Clean solar panels regularly to remove dust, dirt, or snow for optimal performance.

7. Check lead-acid battery water levels if using them, or monitor lithium-ion batteries for performance.

8. Fix any loose connections or signs of wear and tear immediately to avoid system failures.
9. Solar systems can be expanded or combined with wind or hydro systems for even greater efficiency.

Generating power is only part of staying warm. The other part of the battle is making sure you have a suitable shelter and the ability to create fire, which leads us into our next book.

BOOK 5: BUILDING SHELTERS AND STAYING WARM

Chapter 1: Crafting Emergency Shelters

Knowing how to build a shelter isn't just important for wilderness survival scenarios. If you're ever forced to bug out of your home and hit the open road on foot to evacuate from an incoming hurricane or wildfire, for instance, understanding how to build a shelter to keep yourself protected from the rain, snow, or wind as you travel is super important.

The Best Locations for Setting Up Shelters

Where you build your shelter is just as critical as how you build it. A good location can make or break your survival setup. With that in mind, here are some things to keep in mind:

- Avoid low areas like valleys or riverbanks where water might collect during rain.
- Build the shelter near large rocks, hills, or trees to shield your shelter from strong winds.
- Make sure you're near a water source, firewood, and food, but not too close to wild animal trails.
- Stay clear of unstable ground, dead trees (they can fall), and areas that are prone to flooding (such as riverbanks).

When I first started seriously prepping, I learned a lesson the hard way about shelter placement. I was out on a solo trip, trying to be all survivalist and rugged, and I found what seemed like a perfect spot: a nice, flat valley near a stream.

I figured, "Hey, it's got water, it's sheltered, and it looks cozy."

But that night, I got an education I didn't sign up for. It rained. Not just a little drizzle, but a full-on downpour. By morning, the stream I had so happily camped beside had turned into a rushing river, and my entire shelter was submerged. My gear was wet, I was freezing, and I had to spend half the day figuring out how to salvage what I could. That was the moment I realized: *never build in a low area*. Valleys and riverbanks might seem inviting, but they're death traps when it rains!

As I gained more experience, I also started paying attention to wind. On another trip, I set up near a clearing, thinking the view was unbeatable. Big mistake. As soon as the sun dipped below the horizon, the wind picked up like a freight train. My shelter wasn't just buffeted by gusts. It was being blasted from every direction, making the whole night feel like I was trapped in a wind tunnel!

I spent hours trying to anchor it better, but I was freezing and getting frustrated. After that night, I learned to build near natural windbreaks like big boulders or dense tree lines. But even that's not foolproof. One of my worst moments came when I built a shelter too close to what I thought was a "convenient" water source...only to find fresh bear tracks around it the next morning. Thankfully I didn't run into anything.

Project: Build a Lean-To Shelter

Let's get hands-on with a simple, effective shelter design: the lean-to.

Here's what you'll need:

- A sturdy branch or pole for the ridge beam (about 6–8 feet long)
- Several shorter branches for the sides
- Leaves, grass, or bark for covering
- Rope or vines (optional, for securing)

Now follow these steps:

- Find two trees about 6–8 feet apart. They'll hold your ridge beam (If no trees are available, prop one end of the beam on a rock or create a tripod with sticks).
- Place the ridge beam horizontally between the two trees.
- Lean shorter branches at an angle against one side of the ridge beam to create the roof.

  Cover the angled branches with leaves, grass, bark, or anything else available. Layer it thick to block wind and rain.

  Use rope or vines to tie the branches together for added stability. Add rocks or more branches to the base to keep everything secure. If rope, paracord, or vine isn't available, your shoelaces can work in a pinch.

Source

The best thing about the lean-to is how versatile it is, and how effective it is at blocking wind and rain. Take a Saturday or Sunday where you have time off and practice making one. It's fun!

Chapter 2:
Firecraft for Survival

Fire isn't just about staying warm. It's like survival gold. It's your light in the dark, your stove in the wild, and even your rescue signal if things go south.

The Importance of Fire in Survival Situations

Fire is a survival multitool. It gives you heat to stay warm, a way to cook food, and even clean water by boiling it. It can also keep animals at bay and act as a signal for rescuers if you're in a bind.

Plus, let's face it…there's something comforting about the crackle of a fire when everything else feels uncertain.

Imagine being caught in freezing temperatures without fire. Hypothermia can set in fast, especially if you're wet. A fire isn't just a convenience. It's literally your lifeline.

The good news? With a little practice, you can learn to make one almost anywhere.

Methods for Starting Fires Using Friction

Friction-based fire-starting is one of those survival skills that feels like magic when you pull it off. It's not easy at first, but it's incredibly satisfying. Here are a couple of reliable methods:

- **Bow Drill:**

    1. Find a straight stick for the bow and tie a piece of cord or shoelace to both ends.
    2. Use a dry, flat piece of wood as your base (fireboard).
    3. Carve a small notch in the fireboard and place dry tinder beneath it.
    4. Twist a spindle (another stick) into the bowstring, press the spindle into the fireboard, and move the bow back and forth to create friction. Once you see smoke, blow gently on the ember and transfer it to your tinder.

- **Hand Drill:**
    1. Use a dry stick as your drill and a flat piece of wood as your fireboard.
    2. Spin the stick between your palms while pressing it into the fireboard.
    3. Keep at it until you see smoke and an ember forming in the notch.

Using Fire for Cooking, Warmth, and Signal

Once you've got a fire going, it becomes your survival MVP. That's because you can utilize the fire in the following ways:

- Use the fire to cook food safely, especially meat, which can harbor bacteria. You can also roast vegetables, boil water, or even bake bread if you've got the tools.
- Position yourself upwind of the fire to get the most heat. In colder climates, a fire can be the difference between life and frostbite.
- A smoky fire is an excellent way to signal for help. Add green leaves, moss, or damp wood to produce thick smoke that can be seen from miles away.

Project: Build a Long-Term Fire Pit

If you're staying in one place for a while, a well-built fire pit is a game changer. It keeps your fire contained, makes it safer, and improves efficiency.

Here's how to build one:

1. Pick a flat, open area away from trees, brush, and your shelter. Wind can carry sparks, so safety first.
2. Remove all flammable material like leaves and grass within a 5-foot radius.
3. Dig a hole about 1–2 feet deep and 3 feet wide. This helps contain the fire and protects it from wind.
4. Place the rocks around the edges of the pit to act as a heat barrier and keep the fire contained.
5. Dig a small trench leading into the pit to allow airflow. This makes the fire burn hotter and more efficiently.
6. Use a teepee or log cabin structure to get the fire started. Once it's going, you can add larger logs to keep it burning steadily.

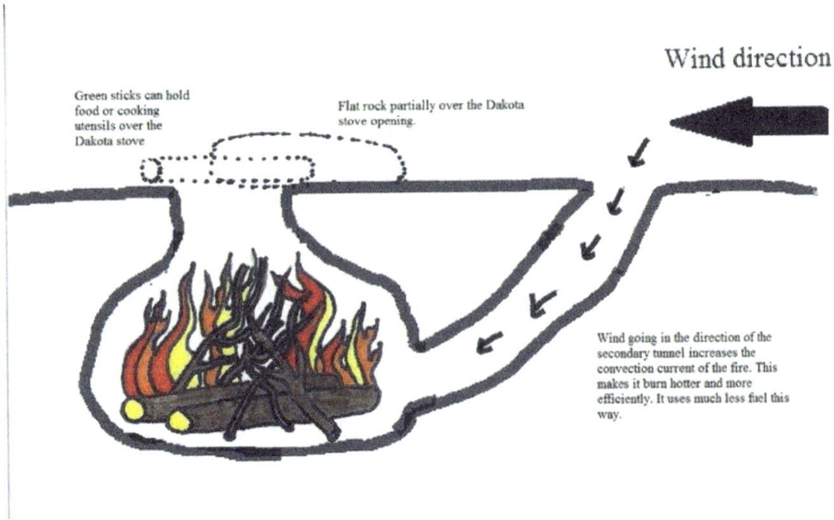

Source

When I built my first long-term fire pit during a camping trip, it completely changed how we handled cooking and staying warm. We picked a flat spot far from trees and cleared out a 5-foot circle of leaves and grass to keep things safe.

After digging a hole about 2 feet deep and 3 feet wide, we lined the edges with rocks to hold in heat and prevent the fire from spreading. My favorite part was adding a small trench leading into the pit and it made the fire burn so much hotter, which meant less smoke and faster cooking. It wasn't fancy, but it was sturdy, efficient, and made us feel like survival pros.

Chapter 3:
Shelter and Heat Management

A poorly insulated shelter or lack of heat can quickly lead to hypothermia, even in mild weather. This is especially true if you're wet.

Let's look at some practical ways to keep your shelter warm and manage heat effectively, even when there's no power.

Insulating Your Shelter for Warmth

A well-insulated shelter can keep you warm without relying on constant heat sources. Here's how:

- Seal the Gaps. Use whatever you have (such as mud, moss, leaves, or even clothing) to block drafts around doors, windows, and walls.
- For natural shelters, pile leaves, grass, or pine needles between layers of branches.
- A cold floor can drain heat fast. That's why you should lay down layers of leaves, straw, or even cardboard to create a barrier between you and the ground.
- Line the inside of your shelter with reflective materials (like aluminum foil or space blankets) to bounce heat back into the space.

Project: Craft a DIY Rocket Stove

Building a rocket stove is a great way to have an efficient, portable heat and cooking source. These small, highly efficient stoves use very little fuel and produce a lot of heat. They're perfect for cooking and warming small spaces. The best part? You can make one from everyday materials:

  - A large metal can (like a coffee can or paint can)
  - A smaller metal can (like a soup can)

- Tin snips or a sturdy knife
- Insulation material (like sand, ash, or small rocks)

Then follow these steps:

- Cut a hole near the base of the large can, just big enough to insert the smaller can. This will be the burn chamber.
- Cut off the bottom of the smaller can and insert it into the hole so it forms an "L" shape with the large can.
- Fill the space around the smaller can inside the large can with your insulation material. This traps heat and directs it upward for maximum efficiency.
- Cut a small hole in the top of the large can to act as the chimney outlet for the heat.
- Place small sticks or biomass into the burn chamber and light them.

And that's it!

Source

Having a solid shelter is a must in any survival situation. It keeps you safe from the weather, helps you stay warm, and gives you a bit of comfort when things get rough. And don't forget, where you build is just as important as what you build. Pick a good spot, and your shelter will work with you, not against you.

Key Points to Remember

1. Use sturdy branches for the frame and pile on leaves or grass for insulation to block wind and rain.
2. Lean-tos and tarp shelters are great for quick protection, while wickiups or log cabins are better for longer stays.
3. Avoid spots that might flood, have loose dirt, or are near dead trees that could fall.
4. Insulate the ground where you'll sleep with leaves, straw, or even cardboard to stay warm.
5. Make sure your shelter is near water, firewood, and food but not too close to animal trails.
6. Secure everything tightly with rope, paracord, vines, or even shoelaces if that's all you've got.
7. Practice building shelters before you actually need one. It's fun and can save you a lot of stress later.

In the next book, we'll spend some time away from home to discuss what you'll need to do to survive in an urban environment.

BOOK 6: URBAN SURVIVAL AND EVACUATION

Chapter 1:
Urban Evacuation Plans

Urban areas can and will get chaotic in a crisis. That's why a solid evacuation plan like we were talking about before isn't just a good idea; it's a must!

Once you know that a hurricane is on its way or an earthquake is likely, you don't want to sit around and evacuate with everyone else. You want to be proactive and take action beforehand to stay ahead of everyone else.

Identifying Escape Routes and Safe Zones

In a city, you can't just wing it when things go south. Escape routes and safe zones need to be figured out ahead of time.

- Take a walk or drive around your neighborhood and map out at least two escape routes. Think about different modes of travel, like walking, biking, or driving. Traffic could be a nightmare, so having alternatives is always a wise idea.
- Where will you go once you're out? Look for places like a relative's house, community centers, or emergency shelters. Make sure they're far enough away from danger but still reachable.
- Major highways or public transit hubs might be packed or chaotic. Stick to less obvious back roads or paths.
- In the heat of the moment, street signs might not help. Memorize landmarks like bridges, parks, or tall buildings to guide you.

Evacuation Timing: When to Leave and When to Stay

Timing is everything in an evacuation. Leave too early, and you risk getting stuck in unnecessary chaos. Leave too late, and you could find yourself in a dangerous situation.

If your home is safe and you have enough supplies, sheltering in place might be the better option. This is especially true if authorities recommend it or roads are completely blocked.

However, you'll need to leave if any of the following occur:

- There's a mandatory evacuation order.
- Your home isn't safe (flooding, structural damage, fire).
- Utilities are out for an extended period, and the situation seems unstable.

Project: Create an Emergency Evacuation Map

A personalized evacuation map is a game-changer. When everything feels chaotic, having a plan laid out can save you time and stress.

Here's how to make one:

1. Print a paper map or download one to your phone that works offline.
2. Highlight your home, safe zones, and potential hazards like flood zones or industrial sites.
3. Include primary and backup routes for walking and driving. Back roads and shortcuts are your friends.
4. Mark landmarks that can guide you in case you get disoriented.
5. Go over the map with your family so everyone knows the plan and meeting points.

When we made our family's emergency evacuation map, it was a total eye-opener. We printed a map of our area and highlighted our house, a few safe zones (like a friend's place and a nearby school), and potential hazards like a nearby river that floods during storms. We mapped out two driving routes and one walking route in case roads were blocked, marking shortcuts and back roads we hadn't thought about before.

Adding landmarks, like a gas station and a big red barn on the way to the school, made it easier for everyone to visualize the plan. Sitting down together to review it as a family gave us all a sense of control, and honestly, a lot more peace of mind.

Chapter 2:
Sourcing Food and Water in Cities

In an urban survival situation, finding food and water can feel overwhelming, but cities have more resources than you might think.

Finding Hidden Water Sources in Urban Areas

There are plenty of sources for water in urban areas:

- Many parks have water fountains or spigots. Check if they're functional and bring a portable water filter to make the water safe.
- Use buckets, tarps, or even large plastic bags to collect rainwater. It's one of the cleanest water sources in a city, but still purifies it before drinking.
- Rooftop air conditioning units often have condensation drip trays. This water isn't potable as-is, but with filtration and purification, it can be drinkable.

Remember to never drink water straight from urban sources without treating it. Always boil, filter, or use purification tablets to remove contaminants.

Urban Foraging: Identifying Edible Plants and Foods

Foraging in a city is a bit different than in the wild, but it's still a valuable skill. Urban areas have plenty of edible plants if you know what to look for:

- Dandelions
    - What to Look For: Bright yellow flowers and toothed leaves.
    - What You Can Eat: Leaves, flowers, and roots.
    - How to Use Them: Toss the leaves in salads or sauté them like spinach. Use the flowers in fritters or teas. Roast the roots to make a coffee substitute.
    - Where to Find Them: Everywhere! Parks, sidewalks, backyards—just avoid areas that might have been sprayed with chemicals.
- Plantains (Not the Banana Kind!)
    - What to Look For: Broad, flat leaves with parallel veins.
    - What You Can Eat: Leaves.
    - How to Use Them: Use raw in salads or cook like greens. Crush leaves for a natural remedy for bug bites or cuts.
    - Where to Find Them: Sidewalk cracks, park edges, and any compacted soil.
- Wild Onions and Garlic
    - What to Look For: Long, thin green shoots that smell like onions or garlic when crushed.
    - What You Can Eat: Greens and bulbs.
    - How to Use Them: Use greens like chives and bulbs like regular onions or garlic.
    - Where to Find Them: Lawns, parks, and green spaces. Be cautious of lookalikes like star-of-Bethlehem.
- Chickweed
    - What to Look For: A low-growing plant with tiny white star-shaped flowers.
    - What You Can Eat: Leaves, stems, and flowers.

  - How to Use Them: Perfect raw in salads or blended into pesto.
  - Where to Find Them: Shady, moist areas in parks or gardens.
- Purslane
  - What to Look For: A low-growing succulent with reddish stems and small green leaves.
  - What You Can Eat: Leaves and stems.
  - How to Use Them: Add to salads, soups, or sauté them. It has a slightly sour, lemony flavor.
  - Where to Find Them: Sidewalk edges, garden beds, or sunny areas.
- Nettles
  - What to Look For: A tall plant with serrated, pointed leaves that sting when touched.
  - What You Can Eat: Leaves.
  - How to Use Them: Boil or steam to neutralize the sting, then use like spinach in soups or teas.
  - Where to Find Them: Near streams, fences, or shady areas. Wear gloves when harvesting!
- Shepherd's Purse
  - What to Look For: A small plant with heart-shaped seed pods.
  - What You Can Eat: Leaves, flowers, and seed pods.
  - How to Use Them: Use leaves raw in salads or cook like spinach. The seed pods add a peppery kick.
  - Where to Find Them: Vacant lots, sidewalks, and garden edges.
- Lamb's Quarters
  - What to Look For: A tall plant with powdery, whitish coating on its leaves.
  - What You Can Eat: Leaves and seeds.

- How to Use Them: Cook like spinach or blend into smoothies. The seeds can be ground into flour.
- Where to Find Them: Gardens, fields, and along fences.

- Wild Mustard
  - What to Look For: Tall plants with small yellow flowers in clusters.
  - What You Can Eat: Leaves, flowers, seeds, and young pods.
  - How to Use Them: Use leaves in salads or stir-fries. The flowers are edible, and the seeds can be made into mustard.
  - Where to Find Them: Vacant lots, roadsides, and parks.
- Mulberries
  - What to Look For: Small trees or shrubs with blackberry-like fruits that can be white, red, or black.
  - What You Can Eat: The berries.
  - How to Use Them: Eat fresh, bake into pies, or make jam.
  - Where to Find Them: Parks, backyards, and urban green spaces.

Scavenging for Supplies in Abandoned Buildings

Abandoned buildings can be goldmines for survival supplies. Here's what to look for:

- Start with the kitchens and bathrooms. These are goldmines. In the kitchen, check cabinets, pantries, and under the sink for canned goods, dry pasta, or cooking tools. In the bathroom, look for bottled water, first aid kits, or even soap and hand sanitizer. Don't forget the back of cupboards because stuff often gets pushed out of sight.
- Bring the right gear. A flashlight or headlamp is a must because these places are usually dark. Wear gloves to avoid cutting yourself on broken glass or sharp edges, and sturdy boots to protect your feet from nails or debris. Bonus points if your boots are waterproof. Flooded basements are no fun.

- Look for containers. Even if you don't find water or food, empty bottles, jars, or pots are super useful for storing things. You can also use them to collect water later or as makeshift cooking tools.
- Be smart about food. Check expiration dates and avoid cans that are dented, bulging, or leaking as they could make you sick. Sealed snacks, rice, or pasta are usually safe, even if they're a little stale. And don't overlook spices and salt. They'll make basic meals way more bearable.
- Think outside the box. Need a rope? Rip up curtains or bedsheets. Looking for insulation? Grab old clothes or blankets. Even things like duct tape, plastic bags, or newspapers can be repurposed into tools or insulation.
- Search closets and storage areas. These often have hidden treasures like coats, boots, or camping gear. Even finding an old tarp or a roll of duct tape could make a big difference in a pinch.
- Don't forget the garage or basement. These are jackpot areas for tools, batteries, rope, or even fuel. Check carefully, though as they can also be the most hazardous areas with rusted metal or puddles of standing water.

Project: Build a Simple Water Filtration System

Clean water is a non-negotiable in any survival situation, and building a basic filtration system is easier than you think.

Here's what you'll need:

- A clean plastic bottle or container
- Gravel or small rocks
- Sand (fine and coarse)
- Activated charcoal (available at pet stores or pharmacies)
- Cloth or coffee filter

Then you can follow these steps:

1. Cut the bottom off the plastic bottle and turn it upside down.
2. Layer the materials inside in this order:
    1. Cloth or coffee filter at the bottom (this will catch fine particles).
    2. Activated charcoal (to remove impurities and odors).
    3. Sand (fine first, then coarse).
    4. Gravel or small rocks at the top.
3. Place the bottle over another clean container to catch the filtered water.

From there, just pour dirty or cloudy water into the top of the filter. Let it pass through each layer slowly. The filtered water should be clearer, but it's still important to boil it before drinking as well to get rid of any harmful substances that you can't see.

When I first built my simple water filtration system, it wasn't because I thought it'd be a fun weekend project. It was a direct response to my growing concern about natural disasters, particularly hurricanes and earthquakes.

Living in a coastal area, I've seen firsthand how quickly things can go south when the water supply gets compromised. I remember one summer, after a particularly nasty hurricane, our city's water was contaminated for days, and everyone was scrambling to find clean drinking water. That experience hit home for me: if the grid goes down and the water stops flowing, you can't just rely on bottled water to get you through. So, I set out to build a simple filtration system for my family.

I grabbed a clean plastic bottle, cut off the bottom, and layered it with gravel, sand, activated charcoal, and a coffee filter like I covered above. The process itself was straightforward, but the real value was in knowing that I had an additional backup plan for clean water.

During the next big storm, when the tap water went out for a few days, we used the filter to clean up water from nearby ponds and streams. While the filter helped clear up the water, I always made sure to boil it before we drank any, just to be safe. It's a simple system, but it works

Chapter 3:
Surviving During Urban Crisis

Urban survival is all about thinking on your feet and being smart with your resources. You need a safe place to stay, a way to protect yourself and your stuff, and a few tech tricks up your sleeve.

Securing Shelter and Staying Hidden in Urban Areas

Finding a solid shelter is priority number one. It doesn't just need to keep you out of the elements, it also needs to keep you safe and off the radar:

- Look for sturdy places like empty warehouses, parking garages, or office buildings. Skip anything that looks sketchy or ready to collapse.
- Keep your spot low-profile. Block light from escaping by covering windows with blankets, tarps, or even garbage bags. Bright lights are like a beacon saying, "Hey, I'm here!"
- Noise is a dead giveaway, especially in the city where sound travels. Keep things quiet. That means no hammering, shouting, or loud music.
- Pick somewhere with multiple ways out. You don't want to be trapped if something goes down.

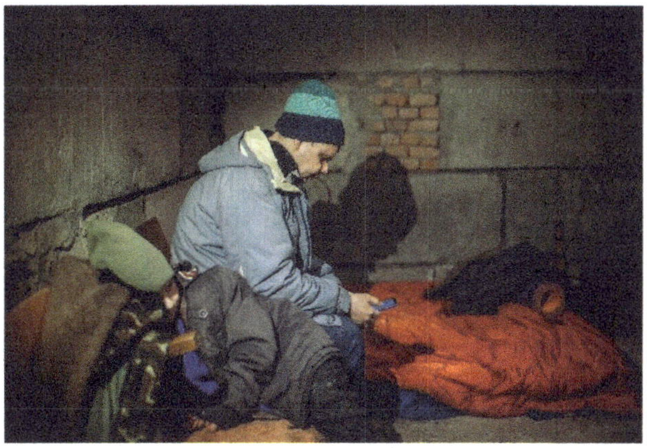

Your goal is simple: to stay invisible. Dress plainly, move with the crowd, and avoid looking like you're carrying anything valuable. The less attention you attract, the better.

Using Technology for Urban Survival: Solar Chargers, Maps, Communication

Technology can be your best friend in a crisis, but only if you use it right. A few simple gadgets can make a huge difference.

- A portable solar charger is a lifesaver. It'll keep your phone, flashlight, and radio running without relying on the grid. Look for one that's foldable and weatherproof.
- Download offline maps of your area. Apps like Maps.me or Google Maps let you navigate without Wi-Fi or cell service. Always have a paper map and compass as backup, because you should know full well, tech can fail.
- A solar or battery-powered radio is great for getting updates on what's happening. Walkie-talkies are also handy if you're with a group and need to stay in touch.

Project: Install a DIY Urban Alarm System

Now let's talk about protecting your shelter with a simple alarm system. It's cheap, easy, and super effective at alerting you to intruders.

Here's what you'll need:

- Fishing line or strong thread
- Aluminum cans or small bells
- Hooks, nails, or tape
- A flashlight (optional for added flair)

Then follow these steps:

1. String the fishing line tightly across doors, windows, or hallways. Make sure it's low enough that it's not obvious but high enough to catch someone.
2. Attach the line to cans filled with rocks or small bells. When the line's disturbed, it'll make noise to warn you.
3. Secure the line with hooks, nails, or tape so it stays in place.

4. Walk through the area and trip the line to make sure it works.

5. Attach a flashlight to the line so it lights up when triggered.

6. Add a second alarm line deeper inside your shelter for an extra layer of security.

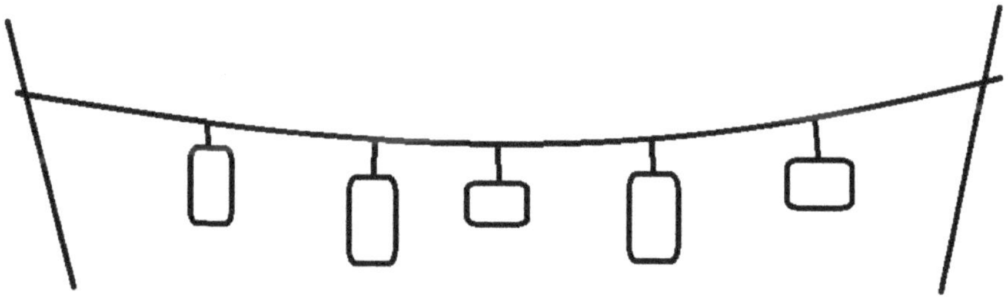

Source

Urban survival is about staying prepared, adaptable, and resourceful when disaster strikes.

I've always prided myself on being a bit of a MacGyver when it comes to prepping. So, when I realized I needed a simple but effective way to protect my shelter from intruders (especially when I'm out in the city during a disaster) I knew it was time to get crafty. I wasn't looking for anything too flashy or expensive; just a basic, no-frills alarm system that could alert me if someone tried to sneak in.

Armed with a fishing line, a few aluminum cans, and some bells, I set to work following the process like I outlined above. I even added a flashlight for a little dramatic flair, so when the line was tripped, the light would flick on and illuminate the area. It felt almost too easy, but honestly, it was perfect for what I needed: a loud, effective, and stealthy alert system that would give me just enough time to react.

The real kicker? I didn't stop at one. I added a second line deeper inside the shelter, so if anyone did manage to get past the first one, they'd still have to face another set of bells (or cans) before getting anywhere near my stash. It wasn't the most advanced system, but it sure as hell got the job done, and it worked perfectly when I tested it out by tripping the line myself.

Key Points to Remember

1. Have a detailed evacuation plan with at least two escape routes and safe zones identified ahead of time.

2. Pack a go-bag with water, food, first aid, protective gear, and tools like multi-tools and flashlights.

3. Know when to evacuate and when to shelter in place based on the situation.

4. Learn to find hidden water sources in urban areas, like park spigots, rainwater, and rooftop condensation. Always purify the water before drinking.

5. Practice urban foraging for plants like dandelions, plantains, and wild onions, and know where to find safe, edible resources.

6. Scavenge abandoned buildings for canned goods, supplies, and tools, but prioritize safety with gloves, boots, and a flashlight.

7. Choose sturdy, inconspicuous shelters and keep your presence hidden by minimizing noise and light.

8. Protect yourself and your supplies with barricades, hidden caches, and simple alarm systems.

9. Use technology like solar chargers, offline maps, and radios to stay powered and informed during crises.

In the next book, we'll return to the home to discuss what you'll need to do to defend your stockpile, possessions, and family should you ever come under attack.

BOOK 7:
HOME SECURITY AND DEFENSE

Chapter 1: Home Fortification

When it comes to home security in a crisis, your goal should be simple: to turn your house into a fortress.

Reinforcing Doors, Windows, and Entry Points

Your doors and windows are the most common entry points, so they're the first things to secure.

- Replace hollow-core doors with solid wood or steel ones. They're much harder to break through. Reinforce the door frame with a strike plate and long screws (at least 3 inches) to keep it securely in place. Adding a deadbolt is a must, but go for one with a reinforced strike box for extra strength.
- Install shatter-resistant window film or security glass to make it harder for someone to break in. Adding window locks or dowels in the tracks can prevent sliding windows from being opened.
- Use a security bar or a piece of wood in the track to stop the door from sliding open. You can also add a sliding door lock or a pin lock for extra security.

Using Landscaping to Create Defensive Barriers

Your yard can do more than just look pretty. It can also act as a first line of defense:

- Plant Defensive Shrubs. Thorny bushes like holly, roses, or barberry under windows make it painful for anyone to try climbing in.
- Trim trees and bushes near entry points to remove hiding spots for intruders. You want a clear line of sight from your home to the perimeter.
- A tall, sturdy fence with a locking gate is an effective barrier. Choose one with vertical slats or wire mesh to make climbing more difficult.
- Gravel around your property makes a lot of noise when walked on, so it can alert you to movement outside.

When I first started thinking about security, I realized my yard could be more than just a place to grill and plant flowers. It could be a key part of my defense plan!

I planted thorny shrubs like holly and barberry under windows to make sure any would-be intruders would regret trying to climb in. It's not a foolproof system, but it sure makes the idea of sneaking into my house a lot less appealing. I also trimmed back trees and bushes near entry points to clear the sightlines. I wanted to see if anyone was lurking around, and those bushes were just giving them cover.

Then, I added a tall fence with vertical slats, which made it a real pain for anyone thinking about scaling it. Plus, gravel around the perimeter made a huge difference. Not only does it look great, but it's surprisingly effective at alerting me to movement because those little stones crunch underfoot.

Installing Steel Bars and Shutters for Added Security

For areas with a higher risk of break-ins, steel bars or shutters can add another layer of protection.

- Installing bars on windows is an effective way to prevent forced entry. Look for bars that can be released from the inside in case of emergencies like fire.
- Rolling shutters are also an excellent option for windows and sliding doors. They can be locked in place to prevent access and provide protection against storms or riots.
- Reinforce your garage door with a security bar or lockable brackets to stop it from being pried open.

Chapter 2:
Surveillance and Monitoring

Keeping an eye on your surroundings is a critical part of home security. Surveillance systems can give you advanced warning of potential threats and help you track activity around your property.

Installing Security Cameras and Motion Detectors

Security cameras are the backbone of any monitoring system, and pairing them with motion detectors makes them even more effective.

- Place cameras at key points like doors, windows, and driveways...basically the main entry points.
- Opt for weatherproof outdoor cameras with night vision and wide-angle lenses. Wireless cameras are easier to install, but wired systems can be more reliable in the long run.
- Motion-activated cameras save storage space and will alert you when there's activity. You can pair them with standalone motion detectors that trigger lights or alarms to discourage intruders.
- Connect your system to your phone or computer so you can receive real-time notifications if something triggers the cameras or motion detectors, such as while you're away at work or grocery shopping.

I'll admit, when I first set up my remote monitoring system, it felt a little like I was living in the future. The idea that I could check in on my property from halfway across town (or even from vacation) was groundbreaking to me, especially during times of crisis or when I'm away for extended periods.

I installed a few modern security cameras that offered cloud storage, which meant I could stream live footage and get alerts directly to my phone whenever something triggered the motion sensors. It gave me peace of mind, knowing I could check in at any time, no matter where I was.

But, of course, I didn't stop there. I made sure to have a local backup, too, using a DVR and an external hard drive to store footage just in case the cloud went down or I lost internet access. I also double-checked that all my cameras were connected to a reliable Wi-Fi signal, and I set up battery backups for the system in case of power outages. It's all about redundancy so I'm not left in the dark when I need it most!

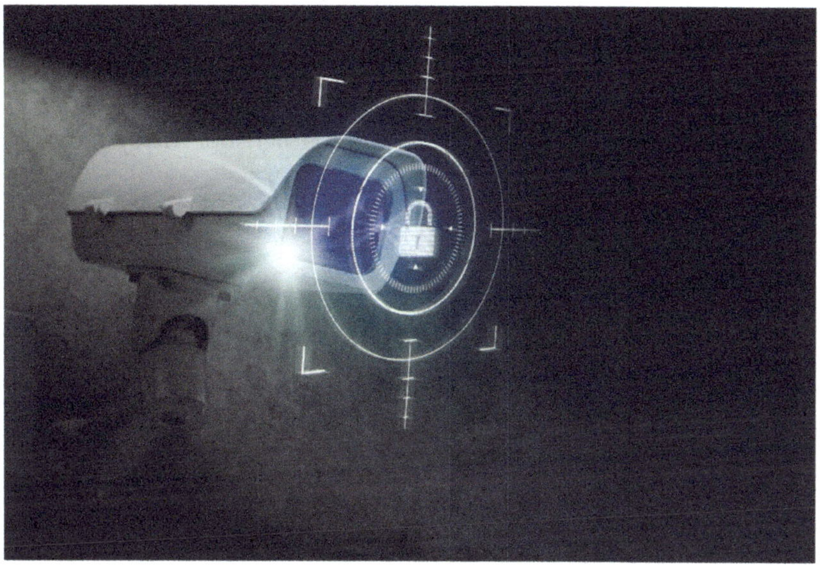

Project: Create a DIY Surveillance Camera Network

Building your own surveillance camera network is easier than you think, and it gives you full control over your home monitoring system.

Here's what you'll need:

- Security cameras (wired or wireless)
- A DVR or cloud storage system for recording footage
- Motion detectors (optional)
- Router and Wi-Fi access for wireless systems
- Mounting brackets, screws, and tools

Then follow these steps:

1. Identify key locations for your cameras, focusing on doors, windows, and other entry points. Don't forget less obvious spots like backyards or alleyways. Mount the cameras securely using brackets and screws.
2. Connect wired cameras to a DVR or wireless cameras to your Wi-Fi network.
3. Link your cameras to a monitoring app or software so you can view live feeds and receive alerts.
4. Test each camera to make sure it has a clear view and no blind spots. Adjust angles and settings as needed.
5. Test motion detectors by walking through their range to ensure they trigger alerts.

Install battery backups or solar-powered options to ensure the system works during power outages.

Chapter 3:
Defensive Tactics for Protection

You don't need to turn your home into a fortress, but having smart strategies, effective weapons, and a few simple early warning systems in place can keep you safe and give you peace of mind in the process.

Non-Lethal Tools for Immediate Protection

Non-lethal tools are effective for deterring intruders and defending yourself without escalating a situation unnecessarily:

- Pepper Spray
  - Compact and easy to use, pepper spray can incapacitate an attacker by causing temporary blindness and intense pain.
- Tasers or Stun Guns
  - These devices deliver an electric shock to temporarily disable an intruder. They're powerful, but you'll need to get close to using them effectively.
- Alarms
  - Personal alarms emit a loud, piercing sound when activated, which will draw attention and potentially scaring off an attacker. These are especially useful for alerting neighbors.
- Flashlights
  - A high-lumen flashlight can temporarily blind someone if it's shined directly into their eyes. Look for a model with a strobe setting, which can be used to temporarily disorient an attacker.

Setting Traps and Early Warning Systems

Simple traps and early warning systems can alert you to potential intruders and give you time to prepare.

- Hang cans filled with rocks or attach bells to tripwires at entry points like doors or windows. These make noise if someone tries to get in.
- Use inexpensive pressure-sensitive alarms under rugs or at entry points. They trigger when stepped on.
- Install motion-activated lights around your property. These not only scare off intruders but also make it easier for you to see what's happening outside.
- Even a "Beware of Dog" sign or fake surveillance camera can act as a deterrent. Most intruders will avoid homes that look like they're well-protected.

One night, my neighbor came over and told me a story that still cracks me up. Apparently, a few days earlier, someone had been trying to sneak around the back of my property, checking windows and doors…real shady stuff.

The guy was getting close to my backyard when he saw the "Beware of Dog" sign I'd hung up near the gate.

Now, let me be clear: I didn't have a dog at the time. But I figured the sign would be a decent deterrent.

Well, it worked. According to my neighbor, the guy stopped dead in his tracks, looked around like he was on the verge of a heart attack, and bolted the other way. No dog, no problem, because the sign was enough to freak him out!

It was one of those small victories that made me realize sometimes the simplest things can be the most effective.

Project: Build a Non-Lethal Defense Kit

A non-lethal defense kit is a portable and effective way to stay prepared for emergencies.

Here's what you'll need:

- Pepper Spray: Look for a compact model that can easily fit in your pocket or bag.

- Taser or Stun Gun: Choose one with a rechargeable battery for long-term reliability.
- Flashlight: A small, high-lumen flashlight with a strobe function.
- Personal Alarm: A lightweight, battery-powered alarm that can be clipped onto clothing or a bag.
- Self-Defense Keychain: Compact tools like kubotans or tactical pens can be highly effective in close-quarters defense.

Then use a small, durable bag or pouch to keep everything organized and accessible. You should store the kit in a location where you can grab it quickly, like near the front door or in your bedroom.

In short, your home is your fortress, and especially in a crisis.

Key Points to Remember

1. Reinforce doors and windows with solid materials, long screws, shatter-resistant film, and additional locks to secure entry points.
2. Use landscaping defensively, plant thorny shrubs under windows, trim bushes to remove hiding spots, and install gravel paths to detect movement.
3. Consider adding steel bars or rolling shutters to windows and sliding doors, especially in high-risk areas.
4. Build a safe room with reinforced walls, a solid door, ventilation, and emergency supplies like water, first aid, and a communication device.
5. Install security cameras at main entry points, paired with motion detectors and real-time alerts for remote monitoring.
6. Use drones to monitor large properties, focusing on hard-to-reach or vulnerable areas.
7. Set up DIY alarm systems with fishing line, cans, or motion-activated lights to detect intruders and give you early warning.
8. Keep a non-lethal defense kit ready, including pepper spray, a taser, a high-lumen flashlight, and a personal alarm.
9. Stay aware of your surroundings, have retreat plans within your home, and use self-defense strategies to keep control in emergencies.

10. Protect your garage and other overlooked entry points with additional locks, reinforced doors, or security bars.

What do you do if anyone in your family gets injured as the result of an attack? That's where knowing first aid and medical skills will come into play, which leads us into our next book...

BOOK 8:

FIRST AID AND MEDICAL PREPAREDNESS

Chapter 1:
Assembling a Medical Survival Kit

When an emergency strikes, having a well-stocked and organized medical kit can mean the difference between a minor incident and a life-threatening situation. You don't need a fancy setup either.

Essential First Aid Supplies for Emergencies

You want your first aid kit to be able to handle everything from minor scrapes to more serious injuries. With that in mind, here's what you should include at the bare minimum::

- Bandages and Dressings:
    - Adhesive bandages (various sizes) for small cuts and scrapes.
    - Sterile gauze pads and rolls for larger wounds.
    - Butterfly closures or Steri-Strips for deeper cuts.
- Antiseptics and Cleaning Supplies:
    - Alcohol wipes or antiseptic wipes for cleaning wounds.
    - Hydrogen peroxide or iodine for disinfecting.
    - Saline solution for rinsing eyes or cleaning wounds.
- Tools and Equipment:
    - Tweezers for removing splinters or debris.
    - Scissors (preferably medical-grade).
    - Thermometer (digital or non-digital).

    - Safety pins and medical tape.
- Medications and Ointments:
    - Pain relievers like ibuprofen or acetaminophen.
    - Antibiotic ointment to prevent infection.
    - Antihistamines for allergic reactions.
    - Anti-diarrheal and antacid tablets.
- Specialized Items:
    - A tourniquet for severe bleeding.
    - Burn cream or gel.
    - Splints or SAM splints for fractures.
    - CPR mask or face shield.

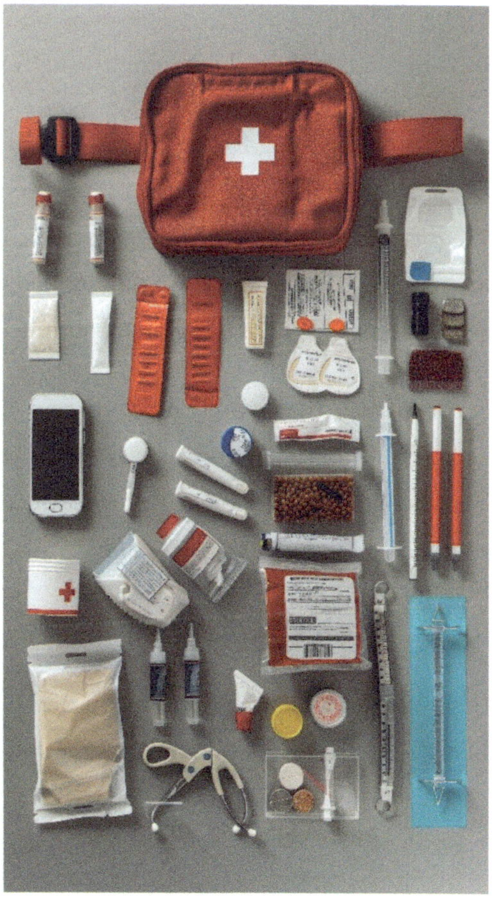

Organizing and Storing Your Medical Kit

A disorganized kit is almost as bad as not having one at all. When every second counts, you need to know exactly where everything is.

- Divide your kit into categories, like wound care, medications, and tools. Use clear plastic bags or labeled pouches to keep everything sorted.
- Use colored tape or markers to make it easy to identify sections at a glance. For example, red for bleeding control, green for medications, and blue for general supplies, and so on.
- Include a checklist in your kit that outlines what's inside. This makes it easy to restock after use.
- Use a sturdy, waterproof case or bag to protect your supplies. Look for something that's portable but also tough enough to handle rough conditions.

I learned the importance of organizing a medical kit sort of the hard way. One day, my son came to me with his first-aid kit. It was a mess of band-aids, ointments, and random tools thrown together in a jumbled heap. I took one look and knew we were in trouble.

"Dad, I can't find the gauze," he said, frantically digging through the chaos.

I watched him waste precious time searching for something simple, and it hit me: a disorganized kit is almost as useless as not having one at all. That's when my wife stepped in. She helped us sort everything out, by creating categories for wound care, medications, and general supplies. We used clear plastic bags and labeled pouches, and she even color-coded everything with tape: red for bleeding control, green for medications, and blue for general supplies.

We also invested in a waterproof bag that would hold up in rough conditions. When my son sprained his ankle a few weeks later, I didn't have to waste any time digging through the kit because I could grab exactly what we needed in seconds. It turns out, when every second counts, organization really is everything!

Project: Create a Portable Medical Kit

A portable medical kit is essential for situations where you need to move quickly while helping someone, regardless of whether it's bugging out or helping someone in the field.

Here's what you'll need:

- A compact, durable bag or case (like a small backpack or tackle box).
- Small containers or pouches for organization.
- Basic first aid supplies: bandages, gauze, antiseptic wipes, and tape.
- Medications: pain relievers, antihistamines, and antibiotics (if available).
- Tools: scissors, tweezers, and a CPR mask.
- Emergency items: tourniquet, gloves, and a small flashlight.

From there, just follow these steps:

1. Pack heavier items at the bottom and lighter items on top for balance.
2. Use pouches or bags to separate categories for easy access.
3. Attach a laminated checklist to the outside so you can quickly see what's inside.

A portable medical kit is your on-the-go lifesaver. That's because it's small enough to take anywhere but packed with enough supplies to handle emergencies.

Chapter 2:
Treating Common Injuries and Ailments

In a survival situation, knowing how to handle common injuries and illnesses can make all the difference. You don't need to be a doctor to be able to provide effective care, which may be contrary to what you think.

Basic First Aid: Cuts, Sprains, Burns, and More

Handling minor injuries properly prevents them from becoming major problems. Here's how to treat some of the most common ones:

- Cuts and Scrapes:
  - Clean the wound thoroughly with water or a saline solution of some sort to remove dirt and debris.
  - Apply an antiseptic, like hydrogen peroxide or iodine, to kill bacteria.
  - Cover the wound with a sterile bandage or gauze to keep it clean and prevent infection.
- Sprains and Strains:
  - Follow the RICE method: Rest, Ice, Compression, Elevation.
  - Wrap the injured area with an elastic bandage for support, but don't make it so tight that it cuts off circulation.
  - Avoid putting weight on the injured area until it's healed up..
- Burns:
  - For minor burns, run cool (not cold) water over the area for 10–15 minutes.
  - Apply an aloe-based gel or burn cream to soothe the skin.

    - For more severe burns (blistering or deep tissue damage), cover loosely with a sterile dressing and seek advanced care if possible.
- Blisters:
    - If intact, leave it alone. Cover with a bandage to protect it.
    - If it pops, clean the area, apply an antiseptic, and cover with gauze.

Recognizing and Treating Shock and Dehydration

Shock and dehydration are two life-threatening conditions that will always need immediate attention:

- Shock:
    - Symptoms: Pale, clammy skin, rapid heartbeat, shallow breathing, confusion, or fainting.
    - Treatment:
        - Lay the person down and elevate their legs to improve blood flow to vital organs.
        - Keep them warm with a blanket or extra clothing.
        - Reassure them and encourage slow, deep breaths.
        - If bleeding is causing shock, apply pressure to the wound and control it immediately.
- Dehydration:
    - Symptoms: Dry mouth, dark urine, dizziness, fatigue, or confusion.
    - Treatment:
        - Give small sips of water or an electrolyte solution frequently.
        - Avoid sugary drinks like soda, as they can worsen dehydration.
        - For severe cases, use oral rehydration salts or make your own by mixing 1/2 teaspoon of salt, 6 teaspoons of sugar, and 1 liter of water.

How to Handle Infections Without Professional Help

Infections can escalate quickly if not treated properly, especially in a survival scenario. Here's how to manage them when you're on your own:

- Clean the infected area with saline or antiseptic solution.
- Apply an antibiotic ointment and cover with a clean, sterile dressing.
- Try warm compresses to improve circulation and help drain pus (if the swelling continues)
- Use over-the-counter pain relievers like ibuprofen to reduce inflammation and fever.

Take note that honey is also a natural antibacterial that can be applied to wounds. Garlic, when crushed and applied topically, has mild antimicrobial properties as well.

I'll be honest, the first time I had to deal with an infection without professional help, I was a bit panicked. It happened during a solo camping trip. It was one of those trips where you think you're prepared for everything... until you realize you're not.

Long story short, I got a nasty cut on my leg while hiking, and before I knew it, the area started swelling, turning red, and looking like it was getting worse instead of better. Normally, I would've run to the nearest doctor, but being miles away from civilization, I had to figure this out myself.

I remembered some basic first-aid training and quickly cleaned the wound with saline (thankfully, I had some in my kit) and applied antibiotic ointment to prevent infection. I covered it with a sterile dressing and tried to keep it as dry as possible. The swelling didn't go down, though, so I used a warm compress to encourage circulation and help drain some of the pus. While I waited for things to improve, I took ibuprofen to reduce the pain and fever.

But the real life-saver?

Honey. I'd heard it was an effective natural antibacterial, so I applied a thin layer to the wound and covered it with a fresh dressing. I also used crushed garlic on the area for its mild antimicrobial properties. It wasn't glamorous, but by the next day, the infection started improving, and I learned just how resourceful you can be when you need to take care of yourself.

Chapter 3: Managing Health in Crisis Situations

When a crisis hits, maintaining your health will always become one of your top priorities. Without access to regular medical care, you'll need to know how to handle chronic illnesses and prevent and manage infections on your own.

Treating Chronic Illnesses Off-Grid

Managing chronic conditions without access to pharmacies or doctors can be challenging, but preparation and creativity can make a big difference.

- If you or a family member rely on prescription medications, talk to your doctor about getting a longer supply. If that's not possible, use automatic refill programs to build a small surplus.
- Research natural or over-the-counter alternatives for your condition. For example:
  - Diabetics can monitor blood sugar levels with test strips and modify diets to manage glucose.
  - High blood pressure can sometimes be reduced through dietary changes like limiting salt and increasing potassium-rich foods like bananas and beans.
- Maintain a written log of symptoms, triggers, and treatments. This can help you identify patterns and make informed decisions.
- Know what to do if your medication runs out. For instance, diabetics should learn the early signs of hyperglycemia and how to adjust their diet and activity to stabilize their blood sugar levels,

Preventing and Managing Infections

Prevention is always your first line of defense against infections:

- Always clean cuts, scrapes, or bites immediately with soap and water or an antiseptic. Cover with a sterile bandage to keep dirt out.
- Wash your hands frequently using soap and water or an alcohol-based hand sanitizer. Make it a habit to clean frequently-touched surfaces regularly.
- Look for redness, swelling, warmth, pus, or fever. Catching infections early gives you a better chance to stop them from spreading.
- To treat infections, you can also use antibiotic ointments for surface wounds. If the swelling or pain worsens, apply warm compresses to encourage drainage.

I remember my cousin getting a nasty cut while we were out hiking, and we learned firsthand how quickly things can go wrong without proper care. Luckily, we had a basic first-aid kit. We cleaned the wound right away with bottled water and an antiseptic wipe, then covered it with a sterile bandage.

Over the next few days, we kept an eye on it for redness or swelling and changed the bandage regularly. When it started looking a little redder than usual, we applied antibiotic ointment and used warm compresses, which helped it heal faster. Catching it early saved us a trip to the ER and kept the situation from escalating.

Herbal Remedies and Natural Medicine for Survival

Herbal remedies can be lifesavers when traditional medicine isn't available. Learning a few key remedies now can help you in a crisis.

- Willow bark contains salicin, which is a natural pain reliever and anti-inflammatory. You can brew it into tea as a substitute for aspirin.
- Ginger and peppermint are excellent for nausea, bloating, and upset stomachs. You can use them as teas or chew small amounts of fresh root.
- Honey has antibacterial properties and can be applied directly to wounds to prevent infection and promote healing. Aloe vera is also great for burns and skin irritations.
- Elderberry and echinacea are known for their immune-boosting properties. You can use them as teas or syrups during cold and flu seasons.

Key Points to Remember:

1. Assemble a well-stocked first aid kit with essentials like bandages, antiseptics, pain relievers, and specialized items like a tourniquet and burn gel.
2. Organize your kit with labeled sections for quick access and keep an inventory checklist to restock items as needed.
3. Regularly review and rotate your medical supplies, replacing expired items and ensuring tools remain in working condition.
4. Learn to treat common injuries like cuts, sprains, burns, and blisters to prevent minor issues from escalating into major problems.
5. Recognize and respond to shock and dehydration by providing proper care and monitoring symptoms early.
6. Understand how to manage infections with cleaning, antibiotics, and natural remedies like honey or garlic when supplies run low.
7. Prepare for chronic illnesses by stockpiling medications, learning alternative treatments, and maintaining health logs for symptom management.

8. Explore herbal remedies like willow bark for pain, ginger for nausea, and aloe vera for burns to supplement medical care when traditional options are unavailable.
9. Create a portable medical kit for emergencies, ensuring it's compact, organized, and equipped with the essentials for on-the-go care.
10. Stockpile medical supplies gradually, focusing on versatile items and storing them in a cool, dry place to extend their shelf life.

In the next book, we'll address how you can stay in touch with your loved ones during an emergency.

BOOK 9: EMERGENCY COMMUNICATION SYSTEMS

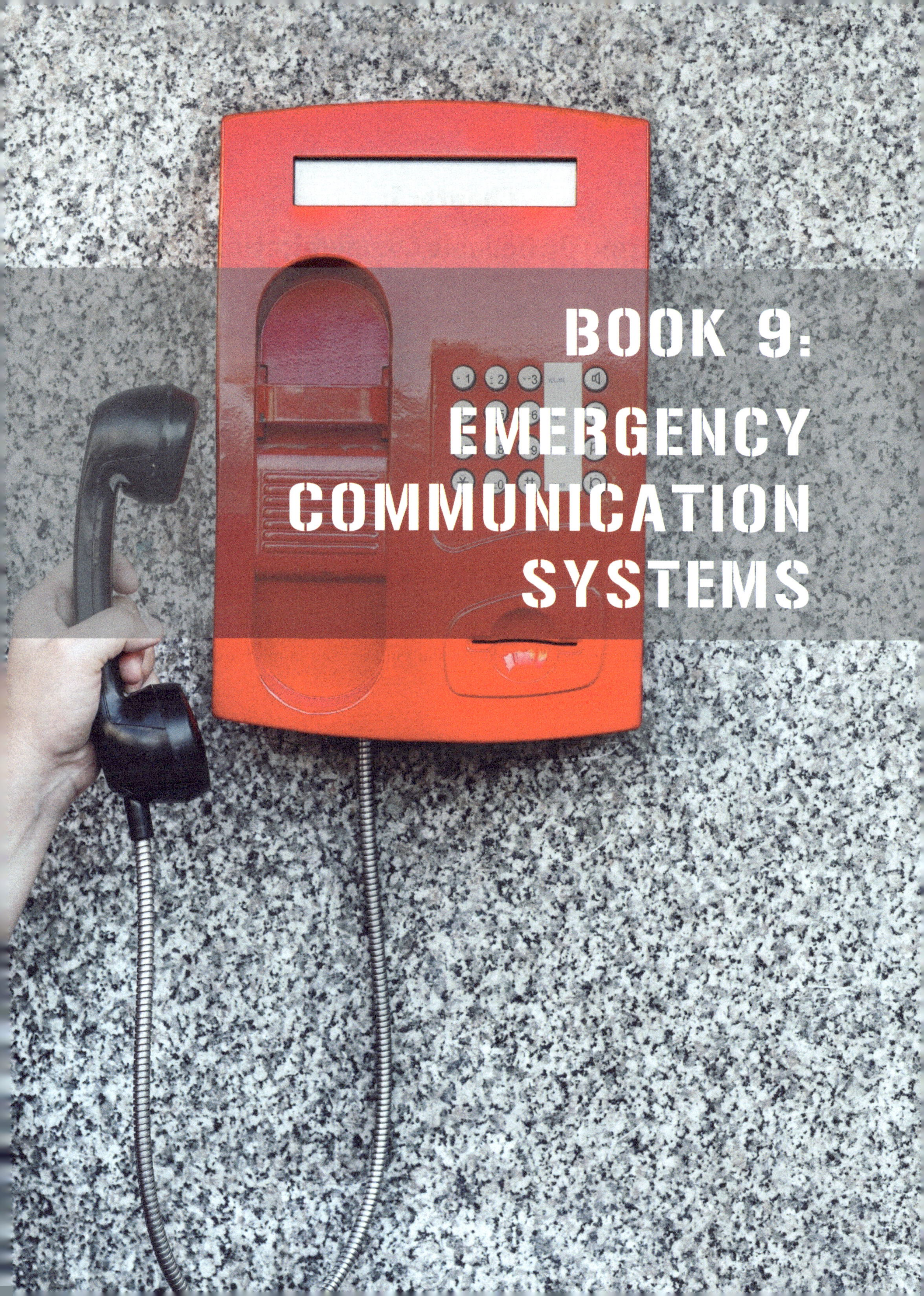

Chapter 1:
Setting Up Reliable Communication

You can't count on cell phones or the internet to work when things go south, so it's smart to have backup communication systems ready. This is especially true for disasters that could knock out the power grid, such as an EMP attack or a natural disaster such as a hurricane or a wildfire.

Ham Radio and Two-Way Radio Systems

If you're serious about emergency communication, ham radios and two-way radios are your go-to tools.

They work even when modern networks fail and can keep you in touch over short and long distances:

- Ham Radios
  - Think of ham radios as the ultimate survival comms. They can connect you to local operators or even people across the world if the conditions are right.
  - You'll need a license to operate a ham radio legally, but the test isn't as hard as you think, and the knowledge you gain is invaluable.
- Two-Way Radios (Walkie-Talkies):
  - Two-way radios are simpler and don't require a license for basic use. They're perfect for short-range communication, like keeping in touch with family or team members during an emergency.
  - Look for models with good range (5 to 10 miles) and rechargeable batteries. Some advanced models even have weather alert functions.

Creating Communication Protocols for Groups

Having the best equipment won't help much if everyone's not on the same page. That's where communication protocols come in. They make sure everyone knows how to communicate clearly and efficiently. With that in mind, here are some tips to remember:

- Decide who in your group will monitor the radio, relay messages, or act as a scribe to log communications.
- Pre-select channels or frequencies for your group to use. Have a primary channel and at least one backup in case of interference.
- Assign call signs (like "Alpha 1" or "Team Bravo") to keep messages clear and anonymous. Avoid using real names over the air.
- Use short, clear phrases. For example: "Alpha 1 to Base, supplies delivered, over." Always end your message with "over" to signal you're done talking.
- Schedule regular check-ins to update each other on status and ensure everyone's okay.

When my friends and I organized a backcountry camping trip, we quickly realized how critical communication protocols were. One person was assigned to monitor the radio at all times, while another logged important updates like our location and supplies.

We agreed on two radio channels: a main one for general chatter and a backup for emergencies. Everyone had a call sign (simple ones like "Team Red" and "Scout 1") so we didn't mix each other up. We stuck to short, clear phrases like "Team Red to Scout 1, firewood collected, over," which avoided confusion.

Regular check-ins every few hours kept everyone accounted for and made sure we were ready if anything unexpected happened. It felt organized, and honestly, it gave us all peace of mind.

Chapter 2:
Emergency Signaling and Navigation

Clear and effective signaling can make all the difference in being found, while strong navigation skills are what will ensure you find safety or resources.

Using Flares, Whistles, and Signal Mirrors

Having the right tools for signaling can make a huge difference:

- Flares
  - Handheld or aerial flares are excellent for drawing attention over long distances.
  - Use them sparingly, as they burn quickly. Aim aerial flares high and away from flammable areas.
- Whistles
  - A whistle is small but powerful. It can be heard from much farther away than shouting.
  - Use three short blasts to signal distress, and then you can repeat that every few minutes to help conserve your energy.
- Signal Mirrors
  - A signal mirror is incredibly effective in daylight. Use it to reflect sunlight toward rescuers, planes, or distant locations.
  - Hold the mirror at eye level and aim the reflected light by sighting through the small hole (if your mirror has one). Move it back and forth to catch attention.

Navigating Using Natural Landmarks

When maps or GPS aren't available, nature can be your guide. Learning to navigate using landmarks and environmental clues can help you find your way to safety.

- Look for landmarks. Mountains, rivers, and large rock formations are excellent fixed points for orienting yourself.
- Rivers and streams often lead to populated areas. If you're lost, follow a water source downhill—it may lead to a town or road.
- The sun rises in the east and sets in the west. If you need a rough idea of direction, observe its movement throughout the day.
- At night, find the North Star (if you're in the Northern Hemisphere). It's located at the end of the Little Dipper's handle and always points north.

Project: Create an Emergency Signaling Kit

An emergency signaling kit should be compact and packed with tools to make you visible and audible when you need help.

Here's what you'll need to build one:

- Flares: Include both handheld and aerial flares if possible.

- Whistle: Choose a loud, durable whistle with no moving parts (pea-less whistles are best).
- Signal Mirror: A lightweight mirror with a sighting hole.
- Flashlight or Headlamp: A small, high-lumen flashlight with a strobe setting is perfect for signaling at night.
- Bright Cloth or Flag: A piece of brightly colored fabric (like orange) can be waved to attract attention.

When I put together my emergency signaling kit, I wasn't just thinking about how I could get someone's attention. I was thinking about how to do it in the most "Hey, I'm in trouble over here!" way possible.

First off, I grabbed a couple of handheld and aerial flares because, honestly, who wouldn't notice a flare shooting up into the sky? Then, I added a whistle, a pea-less one (so it wouldn't get stuck with moisture) because I wanted something loud enough to rival a fire alarm. The signal mirror was a no-brainer. I mean, if you can't reflect the sun into someone's eyes, are you even really trying? I also packed a small, high-lumen flashlight with a strobe setting for nighttime signaling. The idea of flashing SOS into the night sky? Brilliant.

For the finishing touch, I threw in a bright orange piece of cloth. I figured if all else fails, I could wave it around like a madman until someone noticed. The whole kit is compact, portable, and makes me feel like I'm ready for anything…even if it's just to alert a passing hiker that I've been *way* off the trail for too long.

Chapter 3: Maintaining Communication in a Crisis

In any crisis, knowing how to communicate without cell phones or the internet will be a lifesaver.

How to Communicate Without Modern Technology

If cell towers and Wi-Fi go down, you'll need to rely on low-tech but effective methods to get your message across:

- Two-Way Radios (Walkie-Talkies):
    - These are perfect for short-range communication (up to 5–10 miles, depending on the model and terrain).
    - You can use them to coordinate with family or neighbors, and assign specific channels to avoid interference.
- Ham Radios:
    - Ham radios allow long-distance communication, even across states or countries under the right conditions.
    - While they require a license to operate, they're unmatched in a large-scale crisis for staying connected with emergency services and other operators.
- Signal Codes and Symbols:
    - Use universally understood signals, like three blasts of a whistle or flashes of light, to indicate distress.
    - Leave chalk or marker symbols at meeting points to guide others or share information.

- Messengers:
    - In small communities, trusted individuals can relay messages on foot, bike, or vehicle. Establish clear routes and times to ensure consistent communication.

Offline Communication Networks for Communities

Building a communication network for your community can help everyone stay informed and organized together. Here's how you can help to ensure that:

- Distribute two-way radios to key members of your group and establish a primary channel for communication. Use specific times for regular check-ins to share updates and ensure everyone's safe.
- Designate a central location (like a community board or a large tree) where people can leave written messages, updates, or requests for help.
- Create a community map that shows the locations of key resources, safe zones, and vulnerable individuals who may need assistance.

When I first set up a community communication network, it was for one simple reason: I didn't want to be the guy *without* a way to check in on my neighbors when the power went out.

So, I handed out two-way radios to key folks and we set a primary channel for communication.

We scheduled daily check-ins, meaning everyone had to report in at the same time, no excuses. That way, we could make sure everyone was safe and share any important updates.

But radios alone aren't enough, so I also nailed a whiteboard on a tree by the community center as a "message board." Not glamorous, but it worked. People left written notes there for help requests or updates. I even put together a simple map of resources (where to find food, water, first aid, etc.) so if things went south, we'd all know where to go.

It wasn't fancy, but it kept everyone connected and on the same page, even when the power (and internet) was down.

Project: Build a Solar-Powered Emergency Radio

A solar-powered emergency radio will be an invaluable tool for receiving critical information and staying in touch during extended crises when traditional communication sources won't be an option.

Here's what you'll need:

- A small, portable radio (preferably one with AM/FM and shortwave bands).
- A compact solar panel (10W or smaller is sufficient for this project).
- A rechargeable battery pack (to store energy from the solar panel).
- Basic tools: screwdriver, wire cutters, and soldering equipment (if necessary).
- Connecting cables and adapters compatible with your radio and battery pack.

Then connect the solar panel to the rechargeable battery pack. Most panels come with a built-in charge controller to prevent overcharging. You'll also need to attach the battery pack to your portable radio using the appropriate adapter. If your radio has a USB charging port, this part is straightforward.

Test the setup by placing the solar panel in direct sunlight and turning on the radio.

Source

Being able to communicate effectively during a crisis is vital. When cell towers and the internet are down, tools like ham radios, two-way radios, and emergency signaling techniques will keep you connected. It's that simple.

Key Points to Remember

1. Ham radios are excellent for long-distance communication, while two-way radios work best for short-range use and group coordination.
2. Learn and implement communication protocols: assign roles, use call signs, and have primary and backup channels ready.
3. Use emergency signaling tools like flares, whistles, and signal mirrors. The "rule of three" (e.g., three flashes or blasts) signals distress.
4. Navigating without GPS is possible with natural landmarks, rivers, the sun, and the North Star.
5. Build a solar-powered emergency radio to ensure you can receive critical information even during long-term power outages.
6. Establish offline communication networks in your community with message boards, maps, and radio-based check-ins.
7. Keep a portable ham radio kit with a backup power source, external antenna, and a list of key frequencies.
8. Emergency signaling kits should include tools like flares, whistles, bright cloth, and flashlights with strobe functions.
9. Practice using your radios and signaling tools regularly to ensure you're comfortable with them in an emergency.

In the next chapter, we'll dive into greater depth into personal defense methods.

BOOK 10: SELF-DEFENSE AND SECURITY

Chapter 1: Hand-to-Hand Combat

Knowing how to handle yourself in a physical confrontation can literally save your life. You also don't need to be some martial arts master or action hero to defend yourself. It's all about learning a few practical moves that actually work when things get messy.

Basic Techniques for Self-Defense

You don't need to be a martial arts expert to defend yourself. A few simple techniques can go a long way in keeping you safe.

First and foremost, always be alert to your surroundings. Trust your instincts as well. If something feels off, it probably is.

- Focus on Vital Points
    - Aim for areas like the eyes, nose, throat, groin, or knees. Strikes to these points can incapacitate an attacker long enough for you to escape.
    - Use simple but effective strikes:
        - Palm Strike: Drive the heel of your palm upward into the attacker's nose.
        - Knee Strike: Use your knee to hit the groin if the attacker is close.
        - Elbow Strike: If they're within arm's reach, an elbow to the jaw or ribs can be devastating.
- Stay Balanced
    - Keep a strong, stable stance with your feet shoulder-width apart. Bend your knees slightly and stay on the balls of your feet for quick movements.
    - Use your body weight in strikes to maximize impact without overexerting yourself.

Using Evasive Movements to Avoid Injury

Sometimes, the best defense is simply not being where the attacker expects you to be. Evasive movements help you stay out of harm's way:

- Sidestep and Pivot
  - When an attacker lunges or swings at you, step to the side and pivot out of their reach. This disrupts their momentum and gives you a chance to counter or escape.
- Duck and Cover
  - If a punch or object is coming toward your head, duck low and protect your face with your forearms. Keep your movements quick and decisive.
- Use Angles
  - Move diagonally or at an angle rather than straight back. This keeps attackers from cornering you and gives you more options for escape.
- Feign and Redirect
  - Pretend to move in one direction to throw the attacker off, then quickly go the other way. This works well when paired with obstacles to slow them down.

Chapter 2:
Weapon Training for Survival

In a survival situation, having a weapon (and knowing how to use it) can mean the difference between life and death.

Introduction to Survival Weapons: Knives, Bows, and Blunt Objects

Here's a breakdown of the essential survival weapons:

- Knives
    - A knife is one of the most useful tools in survival. It can be used for self-defense, hunting, cutting rope, and more.
    - Learn basic techniques for using a knife in defense, such as slashing (quick and broad) and thrusting (focused and direct).
    - Keep your blade sharp and maintain a secure grip.
- Bows
    - A bow is an excellent choice for both hunting and defense. It allows you to strike from a distance while staying quiet.
    - Practice regularly to improve your aim and learn how to nock, draw, and release efficiently.
- Blunt Objects
    - Tools like bats, crowbars, or even walking sticks can be highly effective in close combat.
    - Focus on targeting vulnerable areas, such as knees, elbows, or the head, to disable an attacker quickly.

Improvised Weapons for Immediate Defense

When traditional weapons aren't available, everyday objects can become effective tools for defense.

Here are some improvised options to give you some ideas:

- Household Items
    - A kitchen knife, hammer, or frying pan can serve as a blunt or sharp weapon in a pinch.
    - A broomstick or mop handle can be used as a staff to create distance between you and an attacker.
- Personal Items
    - Keys can be held between your fingers for close-range defense.
    - A sturdy belt with a heavy buckle can be swung like a flail.
- Environmental Objects
    - Rocks or bricks can be thrown to keep attackers at bay.
    - A heavy branch or fallen limb can be turned into a club with minimal effort.

Project: Build a DIY Self-Defense Weapon

Creating your own self-defense weapon is a valuable skill. That's because it gives you a reliable tool and teaches you to work with what you have.

Here's what you'll need:

- A sturdy wooden handle or branch (around 2–3 feet long).
- Nails or screws (3–4 inches long).
- Duct tape or strong cord.
- Sandpaper (optional, for smoothing the handle).

Then follow these steps:

1. Smooth the branch or handle with sandpaper if you have it.

2. Hammer nails or screws into one end of the handle, and leave about 1–2 inches protruding. Space them evenly.

3. Wrap the other end of the handle with duct tape or cord to create a non-slip grip.

And that's it! Now you can practice swinging the weapon to get a feel for its weight and balance. You can test it on a soft target (like a pillow or old piece of wood) to make sure the spikes are secured as well.

Chapter 3:
Non-Lethal Defense Strategies

Not every situation calls for lethal force…

How to Use Pepper Spray and Tasers Effectively

Pepper spray and tasers are among the most effective non-lethal tools for personal defense. That's because they're easy to use and highly effective at stopping an attacker.

- Pepper Spray
  - Spray directly at the attacker's eyes and nose. The goal is to cause temporary blindness and disorientation.
  - Most pepper sprays have a range of 8–12 feet. Stay at a safe distance and move away as soon as you spray.
  - Familiarize yourself with how your pepper spray works. Practice aiming and spraying to build muscle memory

- Tasers and Stun Guns
  - Tasers work best when the probes hit large muscle groups, like the chest or thighs. For stun guns, aim for areas like the neck or stomach.
  - Tasers fire from a short distance (usually 15–20 feet), while stun guns require direct contact.
  - Deliver the shock, then move away to create distance. Don't linger. It's your chance to escape!

Setting Up Noise Alarms and Other Non-Lethal Defense Tools

Noise alarms and similar devices can scare off intruders and alert others to danger. Here's how to use them:

- Personal Alarms
  - These small devices emit a piercing sound when activated. You can attach one to your bag, keys, or belt for easy access.
- Door and Window Alarms
  - Install battery-powered alarms on your doors and windows. They activate when the entry point is opened, and will then create a loud alert.
- Tripwire Alarms
  - Set up tripwires around your property or campsite. You can attach them to noise-making items like cans or bells, or otherwise pair them with a battery-powered alarm.
- Bright Lights
  - Motion-activated lights not only illuminate dark areas but also startle intruders. Pair them with alarms for a double layer of deterrence.

I'll admit, when I first installed motion-activated lights around my property, I didn't expect them to be so effective.

They seemed simple enough; just a few lights in dark corners, set to turn on when someone walked by. But let me tell you, those lights were *effective.*

One night, I noticed a shadow moving near my garage, and bam, the lights flashed on brighter than a spotlight. The intruder froze, probably more startled than anything, and then bolted. I didn't even need a loud alarm or a big confrontation.

Just the shock of being illuminated like they were on stage was enough to send them running. It's a cheap and easy deterrent, but those lights make it clear: if you're up to no good, you're going to be seen!

Project: Build a Personal Defense Kit

A personal defense kit is a compact collection of non-lethal tools you can carry with you for added safety.

Here's what you could include:

- Pepper Spray: Choose a compact, easy-to-use model with a safety lock to prevent accidental discharge.
- Taser or Stun Gun: Select one with a rechargeable battery and safety features.
- Personal Alarm: Look for one with a loud decibel rating and a long-lasting battery.
- Flashlight: A high-lumen flashlight with a strobe mode for disorienting attackers.
- Self-Defense Keychain: Tools like kubotans or tactical pens are lightweight and effective in close-range situations.

Then you'll want to keep these items around your house and in your vehicle as well.

Protecting yourself and your home during a crisis is all about preparation and having practical strategies.

Key Points to Remember:

1. Learn basic self-defense moves, like targeting vital points (eyes, nose, throat, groin) to incapacitate an attacker quickly.
2. Break free from wrist grabs by twisting toward the thumb, which is the weakest part of their grip.
3. Improvised weapons, such as keys, frying pans, or broomsticks, can be just as effective as traditional ones in emergencies.

4. Pepper spray and tasers are powerful non-lethal tools; aim for the eyes, nose, or large muscle groups for maximum impact.
5. Use tripwires, cans, or bells to create simple early warning systems around your property or campsite.
6. Motion-activated lights and alarms are excellent for deterring intruders and alerting you to threats.
7. Always focus on escaping rather than trying to fight; self-defense is about survival, not winning a fight.
8. Build a personal defense kit with essentials like pepper spray, a flashlight, a taser, and a personal alarm.
9. Practice using your tools and techniques regularly so you're comfortable and confident in an emergency.

In the next book, we'll take a bit of a detour to discuss wilderness navigational skills, which are just as important as personal defense skills but for different reasons.

BOOK 11:

WILDERNESS NAVIGATION SKILLS

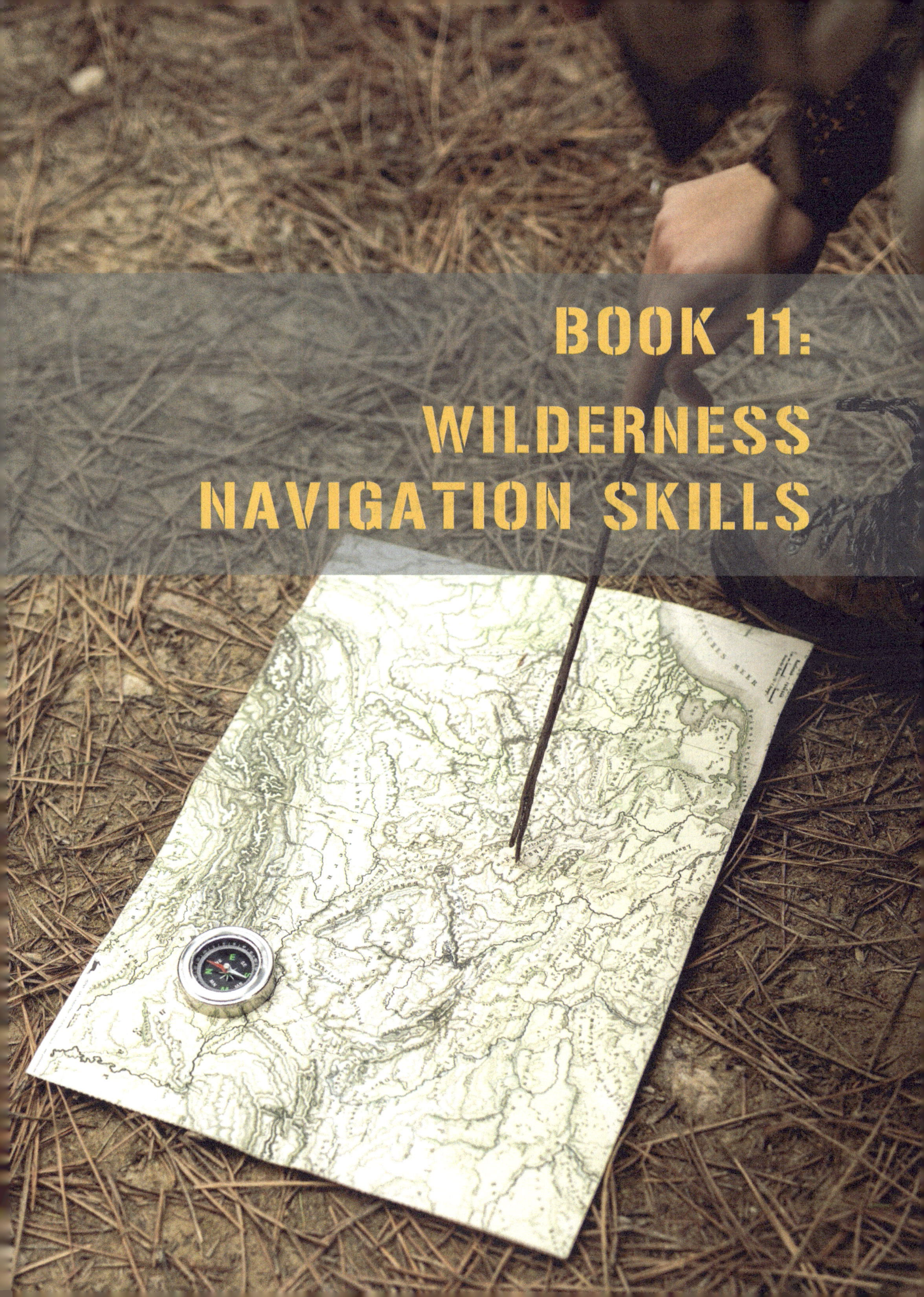

Chapter 1:
Mastering Map and Compass Skills

When you're out in the wilderness, knowing how to navigate without relying on GPS is one of the most vital skills that you can learn. A topographical map and a compass are powerful tools, but only if you know how to use them.

Reading Topographical Maps and Navigating

A topographical map is like your guide to understanding the terrain and planning your route. Here's how to make the most of it:

- Topographical maps use contour lines to show elevation. The closer the lines are, the steeper the terrain. Features like rivers, valleys, ridges, and peaks are easy to spot once you get the hang of reading the map.
- Study the legend to identify key features like trails, source of water, and vegetation types. You should learn to recognize contour intervals (the difference in elevation between lines) to gauge how challenging the terrain will be.
- Identify landmarks along your path, such as hills, streams, or rock formations, to help you stay oriented. Avoid steep climbs and thick vegetation. Instead, look for gradual slopes and clear paths.

Using a Compass to Orient Yourself in the Wild

A compass is an invaluable tool for finding your way. Once you know how to use it, you can confidently navigate even in unfamiliar terrain.

Here are the basic parts of the compass:

  - The needle always points to the magnetic north.
  - The baseplate is the flat part with a ruler for measuring distances on a map.

- The rotating bezel is the circular dial marked with degrees.

Now here's how to use it:

1. Place the compass on the map with the edge pointing from your location to your destination.
2. Rotate the bezel so the orienting lines match the map's north-south grid lines.
3. Remove the compass and turn yourself until the needle aligns with the north arrow on the bezel. The direction of the travel arrow should now point toward your destination.

Project: Build a DIY Navigation Kit

A good navigation kit includes everything you need to stay oriented in the wilderness. Here's what you'll want to remember to include in yours:

- Choose a detailed map of the area you'll be exploring. Waterproof maps are ideal.
- Go with a high-quality compass with a rotating bezel and clear markings.
- Have a ruler and protractor, which are useful for measuring distances and plotting precise bearings on the map.
- Have a pen and notebook for taking notes, marking routes, and jotting down important observations.
- Have a mirror or reflective surface to use as the signal tool or for sighting bearings.

You'll want to gather these items in a small pouch that you can keep in your hiking backpack or in your car so it's always with you.

Chapter 2:
Navigating Without Technology

When technology fails, nature will become your new guide.

Using the Sun and Stars for Navigation

The sun and stars are reliable navigational tools that have guided explorers for centuries.

- Using the Sun
    - The sun rises in the east and sets in the west. By tracking its position in the sky, you can estimate cardinal directions.
    - At noon (when the sun is at its highest point), it will be due south in the Northern Hemisphere and due north in the Southern Hemisphere.
    - To find east-west, stick a straight object like a stick into the ground. Mark the tip of its shadow, wait 15 minutes, and mark the new position of the shadow's tip. Draw a line between the two marks. This is your east-west line.
- Using the Stars
    - In the Northern Hemisphere, locate the North Star (Polaris). It's at the end of the Little Dipper's handle and always points north.
    - In the Southern Hemisphere, find the Southern Cross constellation. Extend an imaginary line along the long axis of the cross to approximate south.
    - Practice identifying these key stars in clear conditions before you need them in an emergency.

I'll never forget the first time I had to navigate by the stars. It was a crisp and clear night while I was out camping, and I realized I had lost my bearings completely. I had no compass and no GPS…just me, the wilderness, and the stars above.

But then I remembered: I could still use the North Star to find my way. In the Northern Hemisphere, it's as simple as locating the Little Dipper and following the handle to the North Star (Polaris). Sure enough, I found it, and there it was pointing directly north. It felt like a small victory, like I had reconnected with something primal.

The Southern Hemisphere had its own little trick up its sleeve, too. I had a chance to try it out on a later trip. The Southern Cross constellation is a gem for orientation. By extending an imaginary line along its long axis, I was able to figure out the general direction of the south. It was a simple trick, but incredibly effective.

Land Navigation Using Natural Landmarks

When the sun and stars aren't visible, natural landmarks can help you find your way:

- Water often flows downhill and may lead to larger bodies of water or human settlements. Follow a stream or river cautiously, and keep an eye on the terrain for easier paths.
- Use mountains as reference points. Peaks are easy to spot and can help you orient yourself if you've seen them on a map or before heading out.
- Moss often grows on the north side of trees in the Northern Hemisphere (and the south side in the Southern Hemisphere). This isn't foolproof, but it can give you a rough direction.

Project: Build a DIY Sun Compass

A sun compass is a simple tool that uses the position of the sun to help you navigate. Here's what you'll need:

- A flat piece of wood or cardboard.
- A straight stick or pencil (for the shadow).
- A ruler and a permanent marker.
- A protractor (optional, for precise angles).

Then follow these steps to build it:

1. Place the stick upright in the center of the flat surface. This will act as the gnomon (shadow-casting object).
2. At noon, mark where the shadow of the stick falls on the surface. Label this mark "South" (Northern Hemisphere) or "North" (Southern Hemisphere).
3. Over the next few hours, mark the shadow's position at regular intervals to create a sun arc.
4. Place your sun compass on level ground. Align the shadow of the stick with the sun arc you created. This will help you determine the cardinal directions.

Source

That's it! A DIY sun compass is a fun tool that you can use to navigate without the aid of technology.

Chapter 3:
Advanced Wilderness Navigation Techniques

Navigating the wilderness like a pro means leveling up your skills. It's that simple.

Backtracking and Wilderness Navigation

Backtracking is all about retracing your steps and recognizing where you've been.

- Break small branches, stack rocks (cairns), or tie ribbons on trees as you move. These little markers will help you find your way back.
- Spot distinctive features like oddly shaped trees, big boulders, or water crossings. These are your signposts.
- Turn around every so often to see what your trail looks like in reverse. It's amazing how different things look going the other way.
- Always take note of where the sun is in relation to your path. If you walked with the sun on your left, it should now be on your right when heading back.

When I got turned around on a solo hike a few years ago, backtracking saved me from spending a very uncomfortable night in the woods. I'd tied bright ribbons on branches at tricky spots and stacked small cairns near forks in the trail, thinking it was overkill at the time.

But when the sun started setting and everything looked the same, those markers were a lifesaver. I also remembered to turn around every so often while hiking, so I already knew what the trail looked like on the way back. By using the sun's position (it had been on my right heading in, so I kept it on my left heading out) I made it back to my car with plenty of daylight left. That little extra effort on the way in made all the difference.

How to Build Makeshift Compasses

When you're stuck without a compass, nature and a bit of creativity can still point you in the right direction.

- Needle and Water Compass:
    - Magnetize a needle by rubbing one end against fabric, hair, or a magnet about 50 times in the same direction.
    - Float the needle on a small piece of leaf, or paper in a cup of still water. It will align with the Earth's magnetic field, therefore pointing north-south.
- Shadow Stick Method:
    - Stick a straight branch upright in the ground on a sunny day.
    - Mark the tip of the stick's shadow with a small rock. Wait 15–20 minutes, and then mark the new shadow tip.
    - Draw a line between the two marks. This gives you an east-west direction (the first mark is west).

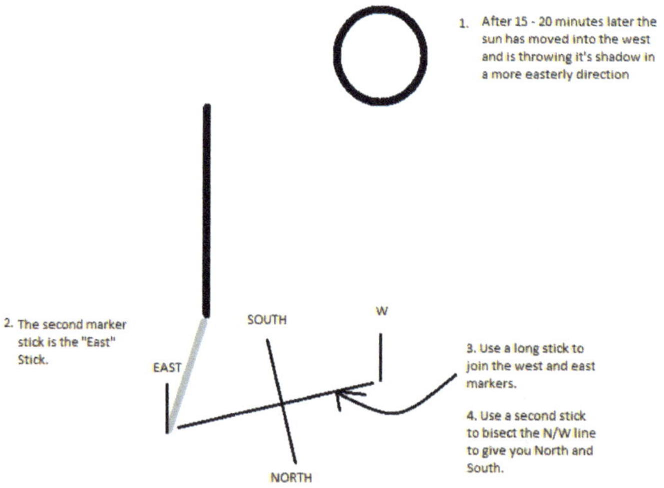

Source

Project: Create a Wilderness Navigation Course

Now it's time to put all your skills to the test! A navigation course is a great way to practice backtracking, using natural landmarks, and building your confidence.

1. Choose a safe area like a park or a familiar trail. Look for features like hills, streams, and clearings to make the course interesting.
2. Create at least five checkpoints using natural landmarks. Mark them on a topographical map and make a list of their descriptions.
3. Include tasks like finding a checkpoint using only a compass, following a backtracking route, or using the shadow stick method to find directions.

After that, just walk the course yourself so you know how everything works.

Being able to use a map, compass, and natural clues can mean the difference between finding your way or getting hopelessly lost. Even without technology, the wilderness itself offers guidance if you know how to read it.

Key Points to Remember:

1. Learn to read topographical maps by understanding contour lines, landmarks, and elevation changes to plan safe routes.
2. A compass is invaluable for navigation. Align it with your map and rotate the bezel to find your direction of travel accurately.
3. Create a navigation kit with essentials like a waterproof map, compass, ruler, and a notebook for quick reference.
4. Use the sun for general direction (rise in the east, set in the west) and midday positioning for north or south.
5. The North Star (Polaris) points north in the Northern Hemisphere, while the Southern Cross guides you in the Southern Hemisphere.
6. Follow natural landmarks like rivers, ridges, and mountains for orientation. Waterways often lead to civilization.
7. Backtracking is critical: mark your trail with small cairns, ribbons, or natural features to help you retrace your steps.
8. Improvised tools, like a needle-and-water compass or shadow stick, can point you north-south when a proper compass isn't available.

In the next book, we'll discuss how you can grow your own crops and food in your home's personal garden.

BOOK 12:

OFF-GRID FARMING AND GARDENING

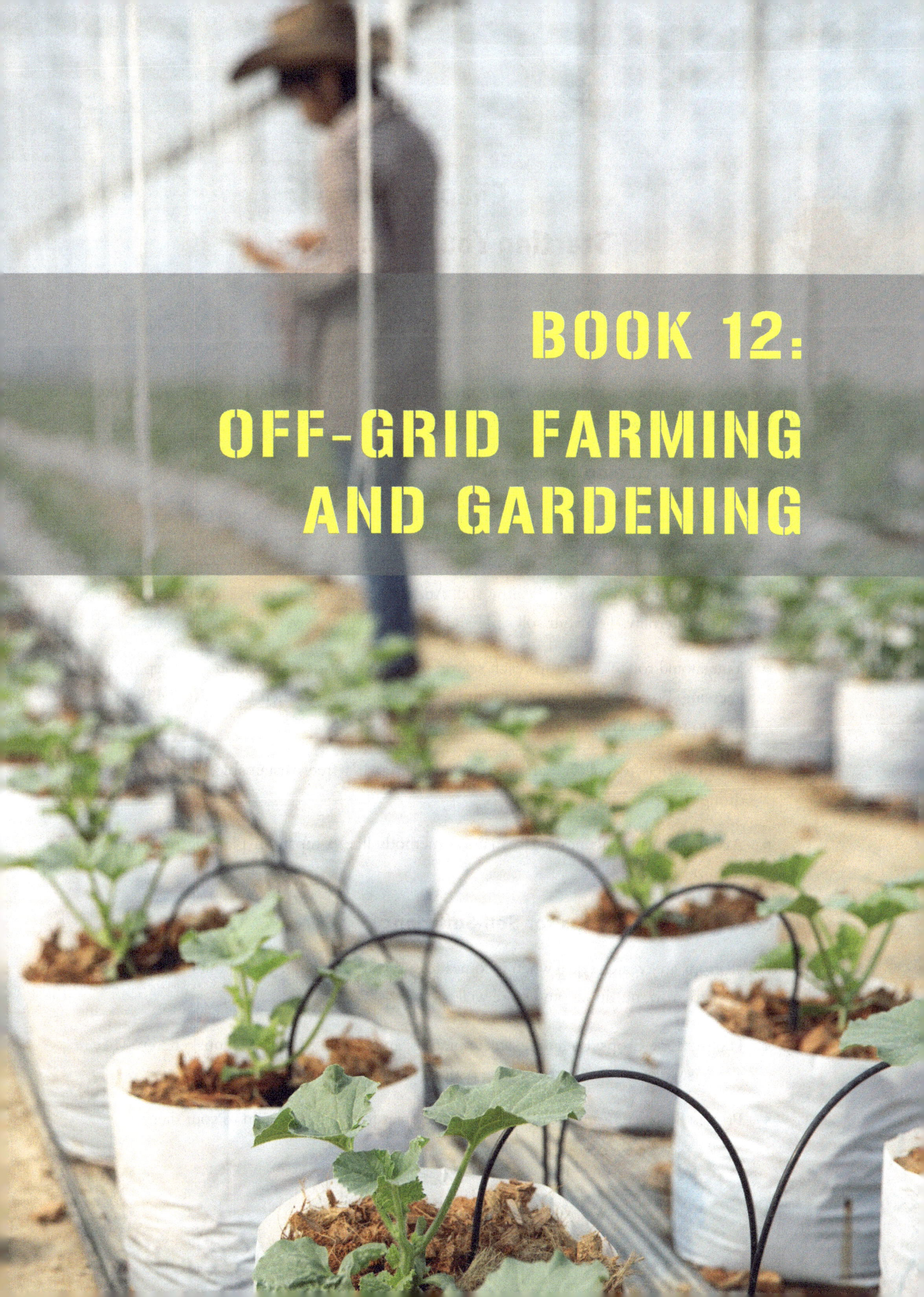

Chapter 1:
Starting Your Homestead

At the core, starting a homestead is about creating a space where you can live sustainably and provide for yourself and your family.

Sustainable Gardening for Survival

Sustainable gardening is all about working with nature, not against it. Here are some key principles to keep in mind here:

- Healthy plants come from healthy soil. Add compost, aged manure, or organic matter to improve fertility. Test your soil's pH and nutrient levels to see what it needs.
- You should rotate your crops each season to prevent nutrient depletion and keep the pests guessing. For example, you can plant leafy greens one year, then legumes (like beans) the next to replenish nitrogen.
- Install barrels to collect rainwater for irrigation. It's free, sustainable, and helps during droughts.
- Avoid synthetic chemicals. Use natural methods like companion planting or homemade fertilizers instead.

Choosing the Best Crops for Self-Sufficiency

Not all crops are created equal when it comes to survival gardening. You should focus on plants that are easy to grow, nutritious, and store well.

- High-Calorie Crops: Potatoes, sweet potatoes, and squash provide essential energy. They're also easy to grow and store for long periods of time..
- Protein Sources: Beans, peas, and lentils are great for adding protein to your diet.

- Leafy Greens: Kale, spinach, and Swiss chard grow quickly and provide essential vitamins. Plus, you can harvest them multiple times a season.
- Herbs: Basil, oregano, thyme, and mint aren't just tasty. They're also medicinal and repel pests.

Project: Start a Small Survival Garden

A survival garden doesn't have to be overwhelming. Start small, and focus on setting up a system that works for you.

1. Find a sunny location with at least 6–8 hours of direct sunlight daily. Make sure that this area has good drainage and easy access to water.
2. Remove weeds and till the soil to loosen it up. Add compost, aged manure, or organic fertilizer to enrich the soil.
3. Start with easy-to-grow plants like lettuce, radishes, carrots, and beans. Add a few potatoes or squash for calorie-dense options.
4. Follow the spacing and depth instructions on the seed packets. Group similar crops together to make watering and care easier.
5. Water consistently, especially during dry spells. Add a layer of mulch (like straw or leaves) to help hold in the moisture and reduce weeds.

When I first started my survival garden, I was all about simplicity: just a few easy-to-grow crops to get started. But as I got deeper into prepping, I realized that the elements can be just as big a threat as pests or poor soil.

Living in a coastal area, hurricanes and heavy rainstorms are always a possibility, so I started thinking about how to protect my little garden. My first step was building a simple windbreak using tall and sturdy plants and a few strategically placed fences. It wasn't a full-on fortification, but it helped to shield my crops from the worst of the wind.

I also added a tarp system to protect against heavy rain. During an unexpected storm, I quickly strung up a tarp over the garden, propped up by a few PVC pipes, to prevent the soil from washing away and to keep the plants from being battered. It wasn't glamorous, but it worked and my garden didn't turn into a mud pit so my seedlings didn't drown.

Chapter 2:
Raising Livestock

Raising livestock is one of the best ways to boost your homestead's self-sufficiency. Animals like chickens, goats, and pigs will always provide you with a steady supply of food.

Caring for Chickens, Goats, and Pigs for Food

Each animal brings something unique to your homestead, and caring for them doesn't have to be complicated.

- Chickens
  - Why Chickens?
    - They're low-maintenance and provide eggs, meat, and natural pest control.
  - What They Need
    - A secure coop for nighttime, fresh water, and a balanced feed diet (along with kitchen scraps and free-range bugs).
  - Care Tips
    - Collect eggs daily to avoid damage or spoilage. Keep the coop clean and dry to prevent disease.
- Goats
  - Why Goats?
    - They're excellent for milk, meat, and even clearing brush on your land.
  - What They Need
    - A sturdy shelter, plenty of hay, and a secure fenced area (they're escape artists).

- Care Tips
  - Trim their hooves regularly and provide mineral supplements like salt licks.
- Pigs
  - Why Pigs?
    - They grow fast, provide a lot of meat, and will eat almost anything, which reduces food waste.
  - What They Need
    - A well-fenced area with plenty of space to root and wallow, plus access to fresh water and balanced feed.
  - Care Tips
    - Rotate their grazing area to prevent overuse and allow the land to recover.

Project: Build a Chicken Coop

A chicken coop is the perfect first project for raising livestock. It doesn't have to be fancy, just functional and secure.

Here's what you'll need:

- Wood (plywood and 2x4s work great)
- Wire mesh (to keep predators out)
- Nails, screws, and a hammer or drill
- Hinges and a latch for the door
- Roofing material (like tin or shingles)

As a golden rule, you'll want to allow at least 2–3 square feet per chicken inside the coop and 8–10 square feet in the run. Choose a spot with good drainage and some shade.

The basic steps for building a chicken coop are as follows:

1. Build a rectangular base and walls from 2x4s. Add a slanted roof for rain runoff.
2. Cover the windows and any openings with wire mesh to keep predators out.
3. Include 1 nesting box for every 3–4 chickens. Line them with straw or wood shavings.
4. Add a sturdy door with hinges and a secure latch.
5. Add perches inside for the chickens to roost at night.

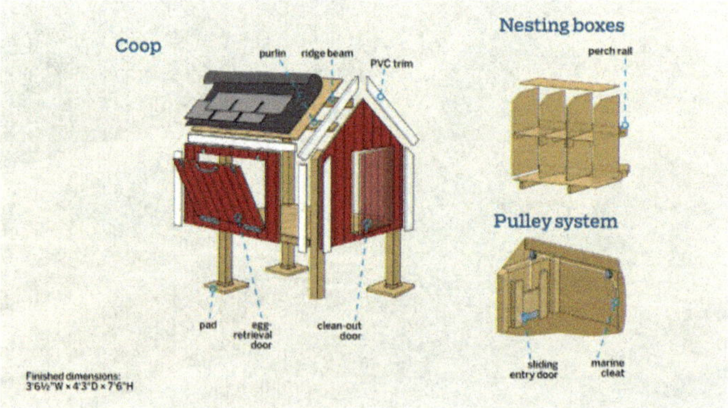

Source

Chapter 3:
Integrating Livestock with Gardening

One of the coolest things about homesteading is how your animals and garden can work together to create a self-supporting system.

Using Livestock Manure to Fertilize Gardens

Manure is nature's fertilizer because it is packed with nutrients that plants love. It's also one of the easiest ways to enrich your soil and boost your garden's productivity.

- Chicken Manure
  - High in nitrogen, chicken manure is a powerhouse for leafy greens and other nitrogen-hungry crops.
  - Compost it first, since fresh chicken manure is too strong and can burn plants. Then allow it to age for 6–12 months or mix it into a hot compost pile.
- Goat and Sheep Manure
  - These manures are milder and can often be applied directly to your garden without composting.
  - They're great for improving soil structure and adding balanced nutrients.
- Pig Manure
  - Pig manure is nutrient-rich but should always be composted due to potential pathogens. A well-maintained compost system will make it safe and effective for your plants.

Project: Build a Simple Greenhouse

A greenhouse is a fantastic addition to your homestead because it will extend your growing season and protect plants from harsh weather. Building one doesn't have to be complicated either. You'll just need the following:

- PVC pipes or wooden frames for the structure.
- Clear plastic sheeting or polycarbonate panels for the cover.
- Zip ties or screws to secure the materials.
- A door or flap for easy access.

Then follow these steps:

1. Use PVC pipes and bend them into an arch or a wooden A-frame to create the structure.
2. Anchor the frame securely in the ground to prevent it from being knocked over by wind.
3. Stretch clear plastic sheeting or polycarbonate panels over the frame. Secure them with zip ties, screws, or clamps.
4. You can install a door or flap to let air in and prevent overheating. For extra airflow, you can cut small vents on the sides that can be opened or closed as needed.

Source

When my neighbor built a simple greenhouse in their backyard, it was a game-changer for their garden. They used PVC pipes for the frame by bending them into an arch and securing the ends into the ground with rebar to keep it sturdy against the wind.

For the cover, they stretched clear plastic sheeting over the structure and held it in place with zip ties and clamps. The final touch was a DIY flap door made from leftover plastic and Velcro, which let them control airflow and temperature. It wasn't fancy, but it kept their plants thriving through an unexpected frost, which proves that you don't need a big budget to protect your garden.

Homesteading is all about self-reliance and sustainability in your everyday life. By growing your food, raising livestock, and integrating smart practices, you create a system that provides for your family while working harmoniously with nature.

Key Points to Remember

1. Improve your soil with compost, rotate crops, and use rainwater for irrigation to maximize your garden's potential.
2. Choose high-yield, self-sufficient crops. Potatoes, beans, leafy greens, and herbs are easy to grow, highly nutritious, and versatile for storage or use.
3. Chickens provide eggs and pest control, goats supply milk and meat, and pigs are excellent for clearing land and producing meat.

4. Regularly inspect livestock for illness, provide clean water, balanced feed, and maintain shelters to ensure healthy animals.

5. Use chickens and ducks for pest control, goats for weeding, and recycle garden scraps as animal feed to minimize waste.

6. Compost chicken and pig manure before applying, but goat manure can often be used directly to enrich soil naturally.

7. Start with projects like a chicken coop or a greenhouse to protect your plants and animals while improving your homestead's efficiency.

8. Use natural pest control methods, crop waste recycling, and seasonal planning to maintain balance on your land.

In the next book, we'll build off of what we've learned about food storage earlier to discuss how you can ensure you always have food over the long term.

BOOK 13: LONG-TERM FOOD SECURITY

Chapter 1:
Growing Your Own Food

When it comes to long-term food security, growing your own food is one of the smartest and most sustainable steps you can take.

Crop Selection for Long-Term Survival

Not all crops are ideal for survival gardening. The best ones are hardy, nutritious, and easy to store as well.

- Calorie-Rich Crops
    - Potatoes and Sweet Potatoes: These are calorie powerhouses and can be stored for months in a cool, dark place.
    - Winter Squash: Butternut, acorn, and other winter squashes store well and are packed with nutrients.
- Protein-Boosting Crops:
    - Beans (Dry and Fresh): They're easy to grow, high in protein, and store well when dried.
    - Peas and Lentils: Great for soil health and providing protein in your diet.
- Nutrient-Packed Greens:
    - Kale and Collards: These are hardy, cold-tolerant, and provide essential vitamins and minerals.
    - Spinach and Swiss Chard: Quick to grow and can be harvested multiple times.
- Grain Alternatives:
    - Corn: Grow flint or dent corn varieties for grinding into flour or meal.

- Quinoa and Amaranth: Both are easy to grow and provide complete protein.

Project: Build Raised Garden Beds

Raised garden beds are a great way to maximize your growing potential. They improve soil drainage and help prevent compaction.

Here's what you'll need:

- Untreated wood planks (2x10 or 2x12 boards work well).
- Screws or nails.
- A drill or hammer.
- Soil and compost to fill the beds.

Then cover these steps:

1. Find a sunny spot that gets at least 6–8 hours of sunlight a day.
2. Cut the wood planks to your desired dimensions. A common size is 4 feet wide by 8 feet long, which allows you to reach the middle from either side.
3. Assemble the planks into a rectangular frame and secure them with screws or nails at each corner.
4. Remove grass and weeds from the area where the bed will sit. Lay down a layer of cardboard or landscape fabric to suppress weeds.
5. Mix the soil, compost, and organic matter to create a rich planting medium. Fill the bed to within a few inches of the top, and you'll be good to go!

Source

Chapter 2:
Stockpiling and Preserving Food

When it comes to preparing for emergencies, stockpiling and preservation are necessary skills that will instantly give you peace of mind.

Building and Storing Your Emergency Food Reserve

Stockpiling food isn't about hoarding; it's about planning. Here are some key tips that you'll want to keep in mind:

- Focus on staples like rice, beans, oats, and flour. These are calorie-dense and versatile.
- Add canned goods like vegetables, fruits, and proteins (tuna, chicken, or beans).
- Include a mix of carbohydrates, proteins, fats, and essential vitamins.
- Don't forget comfort foods like chocolate or coffee. They can boost morale during tough times.
- Use airtight containers to keep out moisture and pests.
- Rotate your stockpile using the "first in, first out" method to ensure nothing goes bad.

When I started building my emergency food reserve, I focused on keeping it practical but also adding a few personal touches. I stocked up on staples like rice and beans because they're affordable, versatile, and store well, and paired them with canned chicken and veggies for balanced meals.

But the game-changer? Adding comfort items like instant coffee and chocolate bars. Trust me, those little treats make all the difference when things get stressful. I stored everything in airtight bins to keep out pests and used sticky notes to track expiration dates so I could rotate items before they went bad. It's not complicated, but it feels good knowing my family is prepared.

Project: Build a Root Cellar for Food Storage

A root cellar is one of the most traditional and effective ways to store food for the long term. It uses the earth's natural insulation to maintain a table temperature throughout the year.

Here's what you'll need:

- A suitable location with well-draining soil.
- Shovels and digging tools.
- Lumber, concrete blocks, or stones for the walls.
- A sturdy door or hatch.

Then follow these steps:

1. Find a location on slightly elevated ground to avoid flooding.
2. Pick an area that stays shaded and cool.
3. Dig a pit that's deep enough to stay below the frost line in your area (usually 6 to ten 10 feet). Create a gentle slope for easy entry or install steps if needed.
4. Line the walls with stone, brick, or wood to prevent soil from collapsing.
5. Add a ceiling with ventilation to regulate humidity and temperature.
6. Add sturdy wooden or metal shelves for storing crates of produce and jars of preserved food.

A root cellar is perfect for storing root vegetables, canned goods, and even fermented foods like sauerkraut.

Growing your food, foraging, and preserving are essential skills for long-term food security.

Key Points to Remember

1. Focus on calorie-dense crops like potatoes, beans, and winter squash, which are nutritious, easy to grow, and store well.
2. Extend your growing season with cold frames, row covers, and hardy perennials like asparagus and rhubarb.

3. Foraging is a valuable skill: learn to identify edible plants like dandelions, wild berries, and mushrooms while knowing the seasons they thrive in.

4. Build raised garden beds to improve soil drainage, reduce weeds, and create a productive growing space.

5. Stockpile essentials like rice, beans, canned goods, and comfort items. Rotate your supplies using the "first in, first out" method to prevent waste.

6. Preserve food through canning, freezing, and drying to maintain a diverse food supply for emergencies.

7. Use natural preservation tools like root cellars to store produce and canned goods at stable temperatures.

8. Livelihood-enhancing projects, such as creating a personalized foraging guide, help you recognize and remember edible plants in your local area.

9. Incorporate a mix of garden and wild foods for a balanced diet that supports self-sufficiency.

In the next book, we'll likewise build upon what we've already learned about water storage to discuss how you can properly conserve and manage your water over the long term.

BOOK 14:

WATER CONSERVATION AND MANAGEMENT

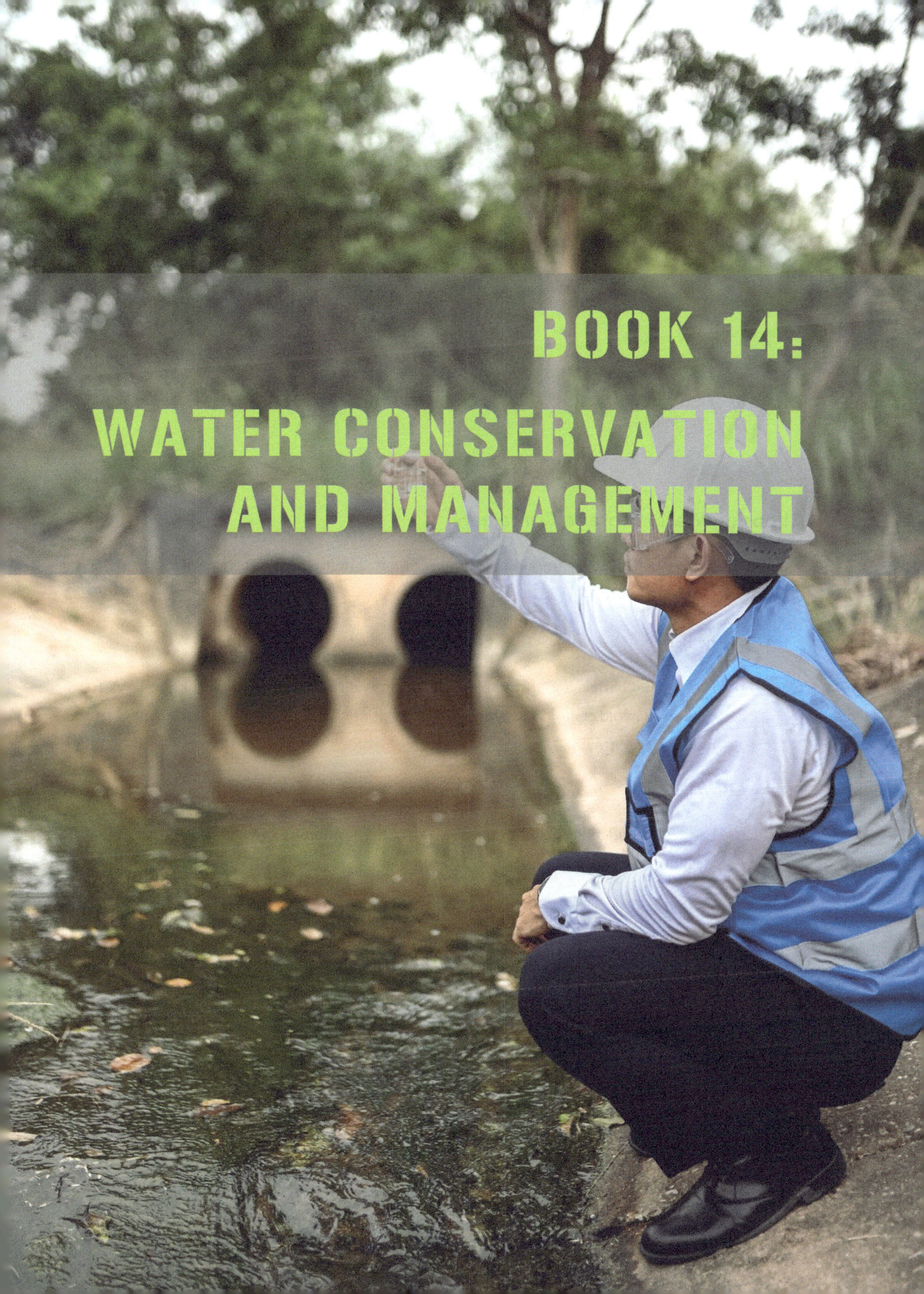

Chapter 1:
Harvesting and Storing Water

Water is life, and in a survival situation, having a dependable water supply is non-negotiable.

Choosing Water Storage Containers

Storing water safely is just as important as collecting it. The right container ensures your water stays clean and ready to use.

- Plastic Barrels
    - Food-grade barrels are affordable and widely available. You should look for BPA-free options to avoid chemical leaching. Blue barrels are best because they block sunlight and reduce algae growth.
- Cisterns
    - Large and underground cisterns are perfect for storing thousands of gallons. They're ideal for long-term water reserves.
- Water Tanks
    - Above-ground tanks made from plastic or metal are durable and come in a variety of sizes.
- Portable Containers
    - Use collapsible water containers or jerry cans for short-term storage or when you need water on the go.

Remember to rotate your stored water every six or so months to ensure freshness.

Project: Install a Rainwater Catchment System

Setting up your own rainwater catchment system is easier than you might think. Here's what you'll need:

- Gutters and downspouts.
- A rain barrel or storage tank.
- A mesh filter to keep out debris.
- A first-flush diverter (optional but recommended).
- A spigot or tap for accessing the stored water.

Then you can follow these steps:

1. Pick a spot under a roof edge where rainwater naturally flows.
2. Attach gutters along the roofline to capture rainwater. Use downspouts to channel the water into your barrel or tank.
3. Place your barrel on a stable platform to allow for gravity-fed water access. Connect the downspout to the barrel, and add a filter to keep out debris.
4. Install a spigot near the bottom of the barrel for easy water access.

And that's it! All you need now is for it to rain.

Chapter 2:
Water Filtration for Off-Grid Living

Filtration removes harmful contaminants, including those that can't be seen, to help make collected water safe to drink.

Building Simple Water Filters with Natural Materials

Nature provides excellent materials for basic water filtration.

- Layered Filters
    - Use a container (like a plastic bottle or bucket) and layer materials in the following order:
        - Top Layer: Coarse gravel (to catch large debris)
        - Middle Layer: Sand (for filtering smaller particles)
        - Bottom Layer: Charcoal)for adsorbing impurities and improving taste)

If you don't have access to pre-made materials, you can gather gravel, fine sand, and burnt wood to create similar filtration layers. Crushed shells or clay can also serve as filtration aids.

When I first tried making a water filter out of natural materials during a camping trip, I was amazed at how effective it was. I used an empty water bottle, cut off the bottom, and layered it with coarse gravel on top, followed by sand, and then some charcoal I made from our campfire. I poured river water through it, and what came out was noticeably clearer. Of course, I still boiled the water afterward just to be safe, but the filter did an incredible job of getting rid of debris and improving the taste. It's a simple trick that can make a huge difference in a pinch.

Project: Build a Gravity-Fed Water Filtration System

A gravity-fed water filtration system is a simple yet effective way to purify large amounts of water with minimal effort.

Here's what you'll need:

- Two large buckets or containers (food-grade plastic is ideal).
- A spigot for dispensing water.
- Gravel, fine sand, and activated charcoal (or regular charcoal if activated is unavailable).
- A drill and a fine-mesh screen or cloth.

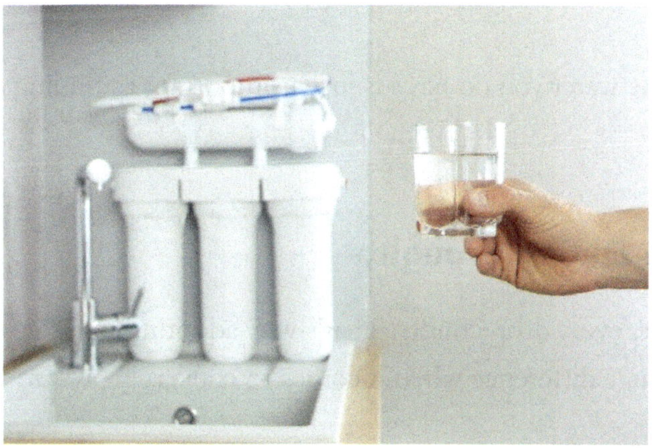

Then follow these steps:

1. Drill a small hole near the bottom of one bucket and attach the spigot securely. This will be the output bucket.
2. Drill a matching hole in the second bucket's base to allow filtered water to flow through to the bottom bucket.
3. In the top bucket, add layers in this order:
    - Gravel (2–3 inches).
    - Sand (4–5 inches).
    - Charcoal (4–5 inches).
4. Place a mesh screen or cloth between each layer to keep materials from mixing.
5. Stack the top bucket (with the filter layers) on top of the bottom bucket.

Now pour water into the top bucket and allow it to filter through!

Chapter 3:
Managing Water Resources

Managing the water you do have is just as important as finding and purifying more of it.

Conserving Water During Droughts

When water is scarce, every drop counts. That's why adopting smart conservation habits like the following can make a huge difference when it comes to conserving water:

- You can maximize your rainwater catchment system by connecting multiple barrels or tanks.
- Capture water from sinks, showers, or laundry to use for irrigation or cleaning.
- Water plants early in the morning or late in the evening to reduce loss from evaporation.
- Mulch your garden to retain soil moisture and reduce the need for you to water it often.
- Fix leaks immediately. Even small leaks in hoses, pipes, or tanks can waste significant amounts of water.

Project: Create a Water Management Plan

A water management plan helps you track your resources and plan for sustainable use. It's a proactive way to stay prepared for emergencies or off-grid living.

1. Identify all available sources of water in your area, such as rainwater, wells, streams, or stored water.
2. Determine your daily water requirements for drinking, cooking, hygiene, and irrigation.
3. Based on that info, you can now assign specific tasks to certain times of the day. For example, you can water the garden early in the morning and collect greywater in the evening.

4. Set aside storage capacity for each source (e.g., rain barrels for rainwater, tanks for well water).

A water management plan is simple when it comes down to it. You need to know how much water you have, how much you're bringing in, and how much you're using up.

Ensuring access to clean water is critical for survival, and it starts with harvesting, purifying, and conserving every drop you can.

Key Points to Remember:

1. Harvest rainwater with gutter systems and food-grade barrels, ensuring safe storage with blue, BPA-free containers to prevent algae growth.
2. Install first-flush diverters in rainwater catchment systems to prevent debris and contaminants from entering your water supply.
3. Use natural materials like sand, gravel, and charcoal to build effective, layered water filters for purifying collected water.
4. Rotate stored water every six months to ensure freshness and avoid contamination.
5. Conserve water by reusing greywater, fixing leaks, and watering plants during cooler parts of the day to reduce evaporation.
6. Adopt efficient irrigation methods like drip systems and grow drought-tolerant crops like beans, squash, and herbs.
7. Build a gravity-fed filtration system for easy, large-scale purification using stacked containers with layered filters.
8. Manage your water usage by prioritizing drinking water, reducing water for hygiene, and using alternatives like hand sanitizers when needed.
9. Mulch your garden to retain soil moisture and reduce how often you need to water.
10. Create a water management plan to track your water sources, storage, and daily usage for sustainable survival.

In the next book, we'll dive into more depth in regards to evacuation plans and ensuring you have suitable shelter should you ever need to bug out.

BOOK 15: EVACUATION STRATEGIES AND SHELTER PLANS

Chapter 1:
Bug-Out Plans and Escape Routes

In an emergency, knowing whether to stay or go can be one of the most critical decisions you'll make.

When to Leave and When to Stay

Deciding whether to hunker down or evacuate depends on the type of emergency and your personal circumstances.

- Stay (Bug-In) If:
    - Your home is secure and has access to essential resources like water, food, and power.
    - The threat is short-term or localized (e.g., minor weather events or temporary power outages).
    - Leaving would expose you to more danger than staying put.
- Leave (Bug-Out) If:
    - There's a high risk to your safety, such as flooding, wildfires, or social unrest.
    - Your home is no longer habitable or secure.
    - The authorities issue mandatory evacuation orders.

Identifying Safe Escape Routes

Not all escape routes are created equal. Knowing your options and planning ahead is key to getting out safely.

- Know Your Local Area

    - Familiarize yourself with major roads, back roads, and alternate routes in your region.
    - Be aware of potential hazards along the way (such as flood zones, bridges, or high-traffic areas).
- Identify Safe Zones
    - Locate destinations like relatives' homes, shelters, or community centers where you can regroup.
    - Choose spots outside the immediate danger area but within a reasonable distance
- Develop Multiple Routes
    - Plan at least two or three escape routes in case your primary path is blocked.
    - Include both vehicular and on-foot routes if you have to abandon your vehicle.

Project: Create an Emergency Evacuation Map

A personalized evacuation map ensures everyone in your household knows how to get to safety.

Here's what you'll want to gather to make one:

- Physical maps of your area (road maps or topographical maps).
- Highlighters, markers, and a notebook for notes.
- A laminator or plastic cover to protect your final map.

Then from there, make sure you do the following on your map:

1. Highlight your home, and nearby safe zones, and potential destinations like shelters, hospitals, or friends' houses. Include gas stations, supply stores, and water sources along the way as well.
2. Draw multiple routes from your home to each safe zone. Use different colors for primary and backup routes. Include notes on footpaths or bike-friendly paths if vehicle travel isn't possible.
3. Mark flood zones, landslide-prone areas, or high-traffic bottlenecks to avoid during an evacuation.

Keep the map in an easily accessible spot, like your emergency kit or vehicle. You'll want to be able to access it and refer to it quickly if you need to!

Chapter 2:
Emergency Evacuation Kits

When disaster strikes, having a well-prepared evacuation kit will be a literal lifesaver.

Essential Items for Bug-Out Bags

So, what exactly goes into a great bug-out bag? You want to focus on the essentials while keeping things light and portable.

Here's what you'll need:

- Water and Hydration: Start with at least 3 liters of water per person. Add a water filter or purification tablets (LifeStraw is a great option).
- Food and Nutrition: Pack high-calorie, non-perishable food like protein bars, jerky, and dried fruit. MREs (Meals Ready-to-Eat) are a solid choice too. Don't forget a lightweight cooking pot or portable stove, utensils, and a can opener.
- Clothing and Shelter: Bring weather-appropriate clothes, including layers and rain gear. Add a compact sleeping bag or an emergency blanket. A small tarp or bivvy sack can double as shelter.
- Tools and Supplies: You'll need a multi-tool or knife (think Swiss Army or Leatherman). A flashlight or headlamp with extra batteries is non-negotiable. Throw in paracord, duct tape, and a small repair kit, Trust us, they'll come in handy.
- First Aid and Hygiene: A good first aid kit is a must. Include bandages, antiseptic wipes, painkillers, and any personal meds. Pack travel-sized hygiene items like soap, toothpaste, a toothbrush, and wipes.

- Navigation and Communication: Grab a physical map of your area and a compass. A whistle for signaling is smart, and a power bank can keep your phone charged for as long as possible.
- Fire and Light: Pack waterproof matches, a lighter, and a fire starter. A few candles or glow sticks can provide light when batteries run out.
- Documents and Cash: Keep copies of your ID, insurance, and emergency contacts in a waterproof pouch. Don't forget some cash in small bills. You never know when ATMs might be out of order.

Project: Build a 72-Hour Emergency Kit

Now it's time to build your own kit. It's easier than you think, and once it's done, you'll feel so much more prepared.

1. Choose a sturdy backpack with plenty of compartments, padded straps, and an adjustable fit.
2. Gather your supplies and make sure you have the basics covered: water, food, clothing, shelter, tools, and first aid.
3. Start with the heavy stuff at the bottom and closest to your back, like water. Middle of the bag is for food and gear, while lighter items like clothing go on top. Use outer pockets for tools and anything you'll need to grab quickly.

4. Once it's packed, wear the bag around for a bit. Is it comfortable? Can you carry it easily?

As a golden rule, you should check your pack every six months and swap anything out accordingly. For example, you can replace food or batteries.

When I built my first 72-hour emergency kit, I started with a backpack I already had from hiking. It had padded straps and plenty of compartments, so it was perfect. I packed it with essentials: water pouches, protein bars, a lightweight tarp, a small first-aid kit, and a multi-tool.

I stacked it carefully, water at the bottom, food and gear in the middle, and clothing on top. Outer pockets held my flashlight, batteries, and a map for quick access.

Once it was ready, I took it for a walk to make sure it wasn't too heavy, and I'm so glad I did because I ended up adjusting how I packed it. Now, I check it every six months to rotate food and supplies, and it gives me so much peace of mind knowing it's ready to grab and go.

Chapter 3:
Surviving While Evacuating

When you're on the move during an evacuation, survival becomes all about adaptability.

Navigating Through Crisis Zones

Crisis zones are unpredictable, and navigating them will require you to be careful:

- Use emergency alerts, radios, or apps to stay updated on road closures, dangerous areas, or evacuation zones.
- Avoid relying solely on GPS. It might not work if the infrastructure is down. Keep a physical map on hand.
- Stick to back roads or trails whenever possible to avoid traffic jams and heavily populated areas. Stay away from high-risk zones like bridges, tunnels, or areas prone to flooding.
- If a route is blocked or becomes dangerous, have backup routes planned ahead of time.

Remember to always trust your instincts. If something feels off, choose another path!

Project: Build a Mobile Shelter for Bug-Out Situations

A mobile shelter is a game-changer for evacuations, providing comfort and protection without slowing you down. Here's how to make one:

Here's what you'll need:

- A sturdy tarp or waterproof fabric.
- Paracord or bungee cords for securing the shelter.

- Lightweight poles or sticks (collapsible tent poles work great).
- Stakes or heavy rocks to anchor the shelter.

Then you can follow these steps:

1. Lay the tarp flat and tie paracord between two trees at shoulder height; you can also use poles if trees are not available
2. Drape the tarp over the cord and secure the corners with stakes or rocks
3. Adjust the sides to create an A-frame design, and anchor the corners and edges with stakes for stability

A simple A-frame shelter like this will be crude, but it will be sufficient for keeping you dry and shielded from the winds.

Having a solid evacuation strategy and shelter plan are super important to your and your family's survival when disaster strikes.

Key Points To Remember:

1. Decide when to evacuate or stay put based on safety, access to resources, and emergency conditions.
2. Plan multiple escape routes with backups, including on-foot options if vehicle travel becomes impossible.
3. Build a personalized evacuation map marking safe zones, alternate routes, and hazards like flood zones or high-traffic areas.
4. Pack a bug-out bag with essentials: water, food, first aid, clothing, navigation tools, and shelter materials.
5. Prioritize efficient packing by placing heavy items close to your back and frequently used gear in outer pockets.
6. Find shelter on the go by utilizing natural features, abandoned structures, or mobile setups like tarp shelters.
7. Navigate crisis zones carefully using physical maps, avoiding high-risk areas, and staying informed via radios or emergency apps.

8. Use a mobile shelter like a tarp-based A-frame to stay dry and protected during emergencies.

BOOK 16: PANDEMIC PREPAREDNESS

Chapter 1:
Stockpiling Medical Supplies for Pandemics

In this chapter we're going to dive into one of the most important aspects of prepping: medical supplies. I'm not talking about a few Band-Aids that are tucked in the back of your junk drawer (most people do that already). I'm talking about being prepared to take care of medical emergencies in a disaster scenario where you don't have access to medical help, regardless of whether it's a pandemic, a natural disaster, or a grid down disaster scenario.

Essential Medications and Medical Tools

When it comes to stocking up on medications (as with any other essentials like food and water), we always want to think long-term. You never know when a pandemic (or, honestly, just a major disaster) might come crashing into your life. And don't just think about the basics like aspirin. You need a full range of meds that could make the difference between you and your neighbor surviving the worst of it.

I would start with the big ones: fever reducers, pain relievers, and allergy medications. Think ibuprofen, acetaminophen, and aspirin. These are your first line of defense when a bug is going around. Stock enough to last at least three to six months. Trust me: running out of these during a crisis is a nightmare.

Next up: antibiotics. This one's tricky because in many places, you need a prescription, but it's worth having a stash of broad-spectrum antibiotics if you can get your hands on them. They can certainly be a lifesaver if you're dealing with an infection and can't get to a doctor.

But don't stop there! We're talking about full-spectrum medical preparedness, so you'll also want to include items like antiseptics, hydrogen peroxide, bandages, gauze, medical tape, and gloves. I keep a few splints on hand too…broken bones are never convenient!

Stocking Food and Hygiene Products for Pandemics

When preparing for a pandemic, food and hygiene products are just as critical as medical supplies. A well-stocked pantry can keep you comfortable and healthy when everyone else is scrambling for the essentials.

Start with non-perishable foods. Like we've covered earlier in this book, canned soups, vegetables, beans, and fruits are great staples to have in your stockpile. Don't forget about grains like rice, pasta, and quinoa. You'll also want some protein sources, so grab a mix of canned meats, tuna, and even protein bars if you're looking for something light but filling.

Now, hygiene. You'll want to stock up on soap, toothpaste, toilet paper, disinfectant wipes, laundry detergent, bleach, and feminine hygiene products. I've been through a situation where the shelves were wiped out, and trust me when I say that being prepared in advance is the way to go.

Project: Build a Pandemic Emergency Kit

Alright, now let's take all the information we've just covered and turn it into a concrete plan. It's time to build your Pandemic Emergency Kit.

Only you can decide what you want to put in your kit, but here's what I would suggest:

1. Medications & Health Supplies:

- Pain & Fever Relief
    - Ibuprofen, Acetaminophen, and Aspirin (at least 3-6 months' worth)
- Prescription Medications
    - Enough for 3-6 months (keep them in their original bottles)
    - Specialty meds (inhalers, insulin, EpiPens, etc.)
- Cold & Flu Medications
    - Cough syrup, decongestants, and throat lozenges
    - Prescription antibiotics (if you have a prescription for them)
- First Aid Kit
    - Bandages, gauze, medical tape, antiseptic wipes, alcohol pads, and hydrogen peroxide

- Splints, burn cream, and antiseptic ointment (Neosporin or similar)
- Tweezers, thermometer, and pulse oximeter

- Personal Hygiene Items
  - Hand sanitizers (alcohol-based, at least 60% alcohol)
  - Soap (bar or liquid, enough for at least 2 months)
  - Disinfectant wipes or sprays

2. Personal Protective Equipment (PPE):

- Face Masks
  - At least 30 N95 or KN95 masks (more if you have a larger family)
- Gloves
  - At least 2-3 boxes of disposable gloves (latex-free if anyone in the family has a latex allergy)
- Eye Protection
  - Safety goggles or face shields (consider if there's risk of airborne illness)

3. Food & Water Supplies:

- Non-Perishable Food
  - Canned soups, beans, vegetables, and meats (3-6 months' worth)
  - Freeze-dried meals (good for long shelf life and easy prep)
  - Rice, pasta, and quinoa (enough to last at least 2 weeks)
  - High-protein snacks like protein bars, nuts, and jerky
  - Granola, dried fruits, and crackers
- Water
  - At least 1 gallon of water per person, per day (for 14 days minimum)
  - Water purification tablets (just in case)

- Baby and Special Diet Supplies
    - Baby formula, baby food, diapers, or anything specific to your family's dietary needs

4. Cleaning & Hygiene Supplies

- Toilet Paper & Paper Towels
    - At least 2-3 months' worth for your family (don't forget to check for shortages!)
- Feminine Hygiene Products
    - Pads, tampons, or menstrual cups (enough to last 2-3 months)
- Disinfectant Supplies
    - Bleach, Lysol, or other disinfectant sprays
    - Cleaning wipes (handy for quick clean-ups)
- Laundry Detergent
    - Enough to last at least 2 months (preferably the concentrated kind to save space)

5. Comfort & Miscellaneous Items

- Entertainment
    - Books, board games, puzzles, and playing cards
    - For kids: coloring books, art supplies, and small toys
- Battery-Powered or Solar-Powered Radio
    - Keep updated on news without draining your phone battery
- Backup Power Supplies
    - Portable power banks for your phone and other electronics
    - Extra batteries for flashlights and other devices
- Paper & Pen
    - In case you need to write down important info (like medical updates or emergency contacts)

Chapter 2:
Quarantine Protocols and Safety

Now that you've got your food and medical supplies stocked up like we just covered, what happens when the illness is inside your house? Or when you need to stay safe while the world is falling apart outside? That's where quarantine protocols come into play. And believe me, it's not just for pandemics. It applies for any kind of disaster where illness could spread, which could happen in the aftermath of a natural disaster (like a hurricane or

Setting Up Home Quarantine Rooms

First things first: you need to have a designated quarantine room (or more than one) inside your house. The goal here is to isolate any family members who may be sick and prevent contamination from spreading throughout the home. It's a simple but effective way to keep the rest of the family safe when one person gets sick.

Here's how to set it up:

- Choose a Room
    - Pick a room that can be easily sealed off from the rest of the house. Ideally, it's a space with a door that can be closed and an attached bathroom (if possible).
- Minimize Shared Surfaces
    - Keep the quarantine room sparsely furnished. A bed, a chair, and a small table are all you need. Remove extra furniture, books, or knick-knacks that could end up getting contaminated and would be difficult to clean.
- Ventilation
    - Proper ventilation in the quarantine room is super critical here. If the room has a window, you'll need to open it if possible to allow fresh air in and reduce the risk of

airborne germs circulating. If not, a fan positioned to push air out of the room can help reduce contamination as well.

I learned the importance of this during the wildfires we faced a few years ago. The smoke was so bad that we had to stay inside for days. We didn't have a true quarantine room, but we did have a safe space where we could close the windows and seal off the rest of the house.

My wife and I took turns staying in the living room, while our kids had the back bedrooms. We managed to minimize exposure to the smoke, and that's when I learned that a simple quarantine space isn't just for pandemics. It's for all sorts of emergencies!

Managing Exposure and Decontamination Procedures

Once your quarantine space is set up, it's time to think about exposure management:

Have a station set up at the entrance of the quarantine room. This station should have gloves, masks, sanitizing wipes, and hand sanitizer. The goal is to avoid carrying contaminants into the space.

Once you've entered the quarantine room (or if you've been exposed to potential contaminants), you'll need to strip down and disinfect. I'm not saying this to be dramatic, but in many situations (like natural disasters or an infectious outbreak) it's a smart idea to wash your clothes immediately and disinfect anything you brought in.

When we had to deal with some flooding a few years back, and I remember how strict we were about washing our clothes and disinfecting ourselves after every trip out to assess the damage. The water was contaminated, and we weren't sure what kind of diseases might be lurking in it. We had a process where we'd come inside, strip down in the garage, and then sanitize everything before even stepping foot in the house. It became second nature, and honestly, it kept us from getting sick.

Project: Create a Home Quarantine Setup

Now that you have a good idea of the steps involved, it's time to create your Home Quarantine Setup. Here's your to-do list:

- Step 1: Choose the Quarantine Space
    - Select a room or area in your home that can be easily isolated. Make sure it's comfortable, has a door, and ideally a bathroom.
- Step 2: Set Up Your Entry Station
    - Place a box or table at the entry of the quarantine room with gloves, masks, hand sanitizer, and disinfecting wipes.
- Step 3: Stock Essential Supplies in the Quarantine Room
    - Make sure the sick person has everything they need (like medications, tissues, a waste bin, water, and snacks).
- Step 4: Plan for Caregiving
    - Designate someone (or rotate the responsibility) to care for the sick family member while taking proper precautions. This person should have personal protective gear ready.
- Step 5: Regularly Disinfect
    - Set a schedule to disinfect surfaces in the sickroom and around the house. I would focus on high-touch areas like door handles and light switches.

Chapter 3:
Treating Illness Without Hospitals

When a crisis hits, hospitals can quickly become overwhelmed, if not completely inaccessible. The good news? You can absolutely do this if you're prepared and know how to manage things.

Managing Respiratory Illnesses and Infections

Respiratory illnesses (whether it's the flu, a cold, or something more severe like pneumonia) can be tricky to manage when medical help is out of reach. However, there are a few key steps you can take to support your loved ones and minimize the impact of respiratory infections.

- Monitor Symptoms
    - If someone in your household shows symptoms of a respiratory illness (fever, cough, shortness of breath), it's super important to monitor their condition closely. For less severe cases, a fever reducer like ibuprofen or acetaminophen can help keep their temperature under control. But be vigilant: if they start having trouble breathing, or if their fever lasts more than a few days or spikes above 103°F, it's time to consider seeking emergency medical help (if and when it becomes possible).
- Maintain Hydration
    - Fever can lead to dehydration, so make sure they're drinking plenty of fluids. Water is your best option, but you can also offer electrolyte drinks or clear broths. Avoid sugary drinks.
- Oxygen Support
    - If the person has severe symptoms like shortness of breath, using a pulse oximeter to check their oxygen levels can help you track their condition. Normal oxygen

saturation levels should be above 95%. If it drops below 92%, it's time to seek medical attention.

I remember a few winters ago when I got hit with the flu. It wasn't just a little cold; it was the full-on, knock-you-out version. I had a fever, aches, and I could barely get out of bed. Luckily, I had already stocked up on supplies in the form of hydration solutions, fever meds, and a pulse oximeter. My wife, who worked as a nurse earlier in her life, kept a close eye on me, and I took full advantage of my "sick-day" to rest and recover. The best thing I did was avoid trying to power through it. Rest is essential.

Using Natural Remedies for Common Illnesses

While modern medicine is an absolute lifesaver, there's something to be said for natural remedies. And in a situation where you're unable to get to a doctor or pharmacy, you'll be glad you've prepared with some tried-and-true natural solutions for common ailments.

- Honey for Coughs and Sore Throats
    - Honey has natural antimicrobial properties and can soothe an irritated throat. For a simple cough or sore throat, a spoonful of honey (especially raw honey) can provide good relief.
- Garlic for Immunity Boosting
    - Garlic is often touted for its immune-boosting properties. You can crush a clove and mix it with honey, or add it to your food for both flavor and health benefits. During the early days of the pandemic, I started throwing garlic into every dish (soup, pasta, you name it) just as a precautionary measure to support immunity.
- Ginger for Digestive Issues
    - Ginger is excellent for nausea and digestive issues. If someone in your family is suffering from an upset stomach, you can make a simple ginger tea by boiling some fresh ginger root in water.
- Echinacea and Elderberry for Immune Support
    - Both Echinacea and elderberry are commonly used to support the immune system. Elderberry syrup, in particular, has become a go-to in our house for warding off colds or flu.

- Saltwater Gargles for Throat Irritations
  - A saltwater gargle can do wonders for sore throats because it helps to reduce inflammation and can kill bacteria. I would mix a teaspoon of salt into warm water and have the person gargle it several times a day.

I've used these remedies with my own family during minor illnesses, like when my daughter came down with a bad cold one winter. She had a persistent cough, and instead of immediately rushing to the pharmacy for a cough suppressant, we tried the honey and ginger tea first. She was much more comfortable, and within a few days, she was on the mend!

Project: Create a Personal Health Log for Family Members

Now, here's the hands-on part. Creating a Personal Health Log for each family member is a great way to keep track of symptoms, medications, and any treatments you've tried.

Here's what I would include:

- Basic Health Information
  - Name, age, medical conditions, allergies, medications, and doctor's contact info.
- Daily Symptom Log
  - Record any symptoms, including fever, cough, shortness of breath, fatigue, and any other symptoms they might experience. Include times of day and severity.
- Medication Log
  - Note all medications being taken (both prescription and over-the-counter), including the dosage, frequency, and any changes.
- Natural Remedies & Treatments
  - Write down any natural remedies you try, their effectiveness, and the person's response.

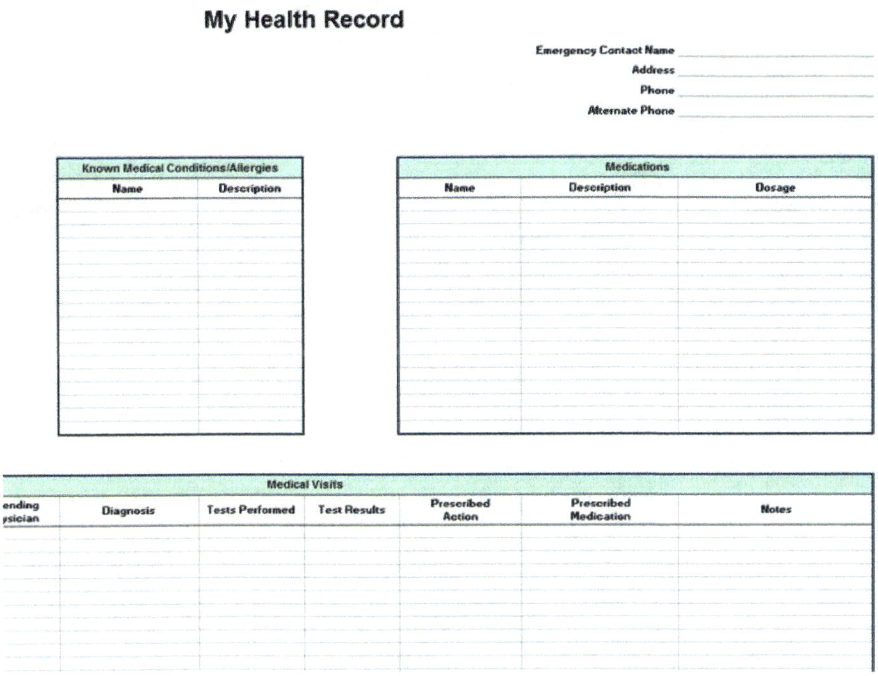

My Health Record

Emergency Contact Name __________
Address __________
Phone __________
Alternate Phone __________

Known Medical Conditions/Allergies	
Name	Description

Medications		
Name	Description	Dosage

Medical Visits						
ending ysician	Diagnosis	Tests Performed	Test Results	Prescribed Action	Prescribed Medication	Notes

<u>Source</u>

Key Points To Remember:

- Stock up on at least three to six months' worth of fever reducers like ibuprofen, acetaminophen, and aspirin.
- If possible, acquire broad-spectrum antibiotics like amoxicillin or ciprofloxacin for emergencies when you can't access a doctor.
- Include medical supplies like bandages, gauze, antiseptic wipes, hydrogen peroxide, medical tape, and splints in your first aid kit.
- Don't forget personal hygiene items like at least two months' worth of soap, disinfectant wipes, toilet paper, and feminine hygiene products.
- For food, stock canned meats, beans, rice, pasta, and protein bars to last at least 3-6 months.
- Plan for at least one gallon of water per person per day for two weeks, plus water purification tablets.

BOOK 17: FOOD PRESERVATION THROUGH FERMENTATION

Chapter 1: Introduction to Fermentation

Fermentation isn't just some trendy food fad. It's actually an ancient preservation technique that's been around for thousands of years.

The Health Benefits of Fermented Foods

Fermented foods are packed with beneficial probiotics, which are good bacteria that can improve digestion, boost your immune system, and even contribute to mental health. You've probably heard of common fermented foods like yogurt, sauerkraut, and kombucha, but what's really happening behind the scenes is that these foods are being "pre-digested" by microbes, which long story short make them much easier for your body to absorb.

When you ferment foods, you increase their nutrient availability and reduce substances that could be hard for your digestive system to handle. For example, the fermentation of cabbage into sauerkraut actually increases the vitamin C content!

I learned firsthand how important gut health is when my family had to quarantine during a flu season. We were relying heavily on canned goods and dry foods, and after a week or so, we started feeling sluggish. So, I turned to fermented foods to help get our digestive systems back in balance. A couple of jars of homemade sauerkraut later, we were feeling much better.

The Science Behind Fermentation

At its core, fermentation is a biological process where microorganisms like bacteria, yeasts, or molds break down sugars and starches in food (which also produces acids, gases, and alcohol). This process preserves food and creates those tangy, complex flavors we love in fermented products.

There are two main types of fermentation:

- Lactic Acid Fermentation

  - This is the process behind foods like yogurt, kimchi, and sauerkraut. Bacteria converts sugars into lactic acid, which not only preserves the food but also makes it tangy and tasty. Lactic acid also helps keep harmful bacteria at bay, which is why fermented foods can last for months without refrigeration.

- Alcoholic Fermentation
  - This is utilized for beer, wine, and some breads, yeast breaks down sugars into alcohol and carbon dioxide. While this is less common for long-term storage, it's still a useful technique to know for certain scenarios, and especially if you're looking to brew your own beverages or make bread rise when yeast is in short supply.

Project: Build a DIY Fermentation Setup

Now that you know the basics, it's time to get started on building your very own DIY fermentation setup. Don't worry, it's simpler than it sounds!

What You'll Need:

- Mason Jars or Fermentation Crock
  - For beginners, mason jars work just fine. If you plan on doing larger batches, a ceramic fermentation crock might be a better choice.
- A Weight
  - You'll need something to weigh down the food so it stays submerged in the brine. You can use a small, clean stone, a jar lid, or a specially designed fermentation weight.
- Salt
  - Use non-iodized salt (like sea salt or kosher salt). Iodine can interfere with the fermentation process.
- Water
  - You'll need purified water for brining.
- Airlock Lid
  - While optional, using an airlock lid for your jar allows gases to escape without letting air in, which can prevent spoilage.

Once you have those items, you can follow these steps:

1. Start with something simple, like cabbage for sauerkraut or cucumbers for pickles.
2. For sauerkraut, shred the cabbage. For pickles, cut the cucumbers into spears or leave them whole.
3. Sprinkle salt over the veggies and let them sit for about 10-15 minutes. The salt will help draw out moisture, which is crucial for creating the brine.
4. Pack the salted vegetables tightly into the mason jar, leaving about an inch of space at the top. Pour the brine (water and salt mixture) over the vegetables to cover them completely.
5. Place your weight (a clean stone or fermentation weight) on top of the vegetables to ensure they stay submerged in the brine.
6. Use a regular mason jar lid or a specialized fermentation lid if you have one. Leave the jar in a cool, dark place for 3-7 days (depending on the temperature and your taste preference).
7. After a few days, start tasting your ferment. When it reaches the level of tanginess you prefer, place it in the fridge or store it in a cool, dark place for long-term storage.

Chapter 2:
Fermentation Equipment and Techniques

Fermentation can seem like a mysterious process, but with the right equipment and techniques, you'll be fermenting like a pro in no time.

Essential Fermentation Tools and Containers

To get started with fermentation, you don't need to go out and buy a bunch of specialty items. Some basic equipment will do the trick, but as you get more into fermenting, you might want to add a few additional items to make the process easier and more efficient.

Here's a list of the most essential fermentation tools:

- Mason Jars
    - The most common and easy-to-find container for fermentation. They're great for smaller batches of fermented vegetables, fruits, or even dairy.
- Fermentation Crock
    - If you want to go big and ferment larger batches of food (or just love the traditional look of a ceramic crock), then you'll want a fermentation crock. These typically come with weights to keep the vegetables submerged in brine and have a lid that lets gases escape while keeping the outside air, well, out.
- Weights
    - Keeping food submerged in brine is crucial to a successful ferment. A weight prevents the vegetables from floating above the brine, which can lead to spoilage. I would go with fermentation weights (yes, it's a thing!).
- Airlock Lids

   - Airlocks are optional, but they're great for letting gases escape while preventing air from getting in. If you're using mason jars, you can buy an airlock lid to fit the jar, which allows gases to bubble out without letting air in (this reduces the chance of mold forming).
- Slicing or Shredding Tools
   - You'll want a sharp knife or a mandolin slicer to cut up your vegetables or fruits before fermentation. A mandolin can save you time if you're making larger batches, but a regular knife will work just fine for smaller amounts.
- Mixing Bowls
   - For most fermentation projects, you'll need a bowl to mix ingredients and massage salt into the veggies or fruits. Stainless steel or glass bowls are perfect because they won't react with the fermenting process.
- Cloth or Cheesecloth
   - If you're using a large crock without a lid, you'll need a cloth or cheesecloth to cover the top of your ferment while it's resting.

Preparing Vegetables, Fruits, and Dairy for Fermentation

Fermentation is as much about preparation as it is about patience. Properly preparing your ingredients will ensure a successful ferment and will also help to improve the overall flavor.

- Vegetables

  - For vegetable ferments like sauerkraut, pickles, or kimchi, start by washing and peeling the vegetables. Cut them into uniform sizes by shredding cabbage for sauerkraut or slicing cucumbers for pickles. Don't forget to massage salt into the vegetables to help draw out moisture and create the brine as well.

- Fruits
  - Fruits can be fermented, but they require a different approach than vegetables. Many fruits (like apples or pears) can be fermented into tangy treats. Start by peeling and cutting the fruit into bite-sized pieces. You might need to add sugar or other sweeteners, as they help with the fermentation process by feeding the natural yeasts in the fruit.

- Dairy
  - Dairy fermenting is a whole different world. If you're looking to make yogurt or kefir, you'll need starter cultures, which are specific strains of bacteria and yeast that help create the fermentation. Once you heat your milk to the right temperature and add the starter culture, you'll need to keep it warm to encourage fermentation.

In one of my early fermenting experiments, I made my first batch of homemade yogurt using just milk and a yogurt starter culture. I didn't have a fancy yogurt maker, so I used a makeshift solution in the form of a towel-wrapped cooler. That batch of yogurt was the creamiest I've ever tasted.

Project: Craft a Fermentation Crock from Recycled Materials

Here's a fun and useful project that will get you thinking outside the box: crafting your own fermentation crock from recycled materials is a great way to save money and get more prepared.

What You'll Need:

- A Large Glass Jar or Ceramic Container
  - Find a large and food-safe glass jar or ceramic container. You can often find these at thrift stores or online.
- A Lid or Heavy Weight
  - You'll need something to keep the food submerged.
- A Cloth or Cheesecloth

    - This will cover the crock to keep dust and debris out while allowing air to flow.
- Non-Iodized Salt
    - Make sure to use the right salt for fermenting, as iodized salt can interfere with the process.

Once you have the above items, follow these steps:

1. Follow the same steps as you would for any fermentation project: wash, cut, and salt your vegetables or fruits.
2. Fill your container with the prepared food, leaving about an inch of space at the top.
3. Place a clean weight on top of the food to keep it submerged in the brine. Then you can cover the opening of the container with cheesecloth or a cloth to protect it from debris.
4. Store the container in a cool and dark place and let it ferment for up to a week.

Chapter 3:
Recipes and Preservation Ideas

Fermentation is as much about creativity as it is about practicality.

Fermented Vegetables, Sauerkraut, and Kimchi

Fermented vegetables are the foundation of many prepping pantries, and the good news is they're easy to make.

Sauerkraut Recipe:

Ingredients:

- 1 medium head of cabbage (about 2 pounds should be sufficient)
- 1-2 tablespoons of non-iodized salt
- Optional: caraway seeds, juniper berries, or garlic for flavor

Instructions:

1. Remove the outer leaves of the cabbage, and then cut it into quarters and remove the core. Slice the cabbage into thin strips.
2. Place the shredded cabbage into a large bowl and sprinkle it with salt. Massage the cabbage with your hands for about 10-15 minutes, until it starts releasing liquid.
3. Pack the cabbage tightly into a clean jar, pressing it down as you go to release more liquid. You want the cabbage to be fully submerged under the brine. If necessary, use a fermentation weight or a smaller jar to keep it submerged.

4. Cover the jar with a cloth or loose lid and leave it at room temperature for three to seven days. Taste it every day; once it's reached your desired tanginess, you can move it to the fridge to slow down the fermentation process.

Kimchi Recipe

Kimchi is a spicy Korean fermented vegetable dish that's typically made with napa cabbage, radishes, and a blend of chili paste, garlic, ginger, and fish sauce.

Here's what you'll need:

- 1 medium head napa cabbage
- 3 tablespoons of sea salt
- 1 tablespoon of grated ginger
- 1 tablespoon of garlic, minced
- 2 tablespoons of fish sauce
- 2 tablespoons of sugar
- 2 tablespoons of Korean red chili flakes (gochugaru)
- 1 small daikon radish, julienned (optional)

Then you can follow these steps:

1. Cut the cabbage into quarters, remove the core, and then chop it down into bite-sized pieces. Massage the salt into the cabbage and let it sit in a bowl for 2 to 3 hours to draw out moisture.
2. In a small bowl, mix together the ginger, garlic, fish sauce, sugar, and chili flakes to form a paste.
3. Once the cabbage is wilted, rinse it to remove excess salt. Mix the cabbage with the radish and the chili paste.
4. Pack the kimchi into a clean jar, and press down to ensure the vegetables are submerged in their own juice.
5. Cover loosely with a lid or cloth and leave at room temperature for up to a week. Taste the kimchi daily, and once it's tangy enough for you, you can store it in the fridge.

Making Kombucha and Kefir

When it comes to fermented beverages, kombucha and kefir are two of the most popular choices.

Kombucha Recipe

Kombucha is a fermented tea that has a slightly tangy flavor. It's made with tea, sugar, and a SCOBY (Symbiotic Culture of Bacteria and Yeast), which is the magic that starts the fermentation process.

You'll need these items:

- 4 cups of water
- 2-3 tablespoons of loose tea or 2 tea bags (black or green tea)
- 1 cup of sugar (white sugar works best)
- 1 SCOBY (you can order one online or find it from another kombucha maker)
- 1 cup of pre-made kombucha (from a previous batch or store-bought)

Then you can follow these steps:

1. Boil 4 cups of water, add the tea, and steep for 5 to 10 minutes. Then remove the tea bags or strain the leaves and stir in the sugar until dissolved.
2. Allow the sweetened tea to cool at room temperature.
3. Pour the cooled tea into a clean glass jar, then add the SCOBY and the pre-made kombucha.
4. Cover the jar with a cloth or paper towel and secure it with a rubber band. Let it ferment at room temperature for up to two weeks, depending on how strong and tangy you want your kombucha to be. Taste it every few days.
5. Once it reaches the desired flavor, you can remove the SCOBY and reserve it for your next batch. Then bottle up the kombucha and store it in the fridge and you'll be good to go!

Kefir Recipe

Kefir is a fermented dairy drink, somewhat similar to yogurt but much thinner and more drinkable. It's made by fermenting milk with kefir grains (a combination of bacteria and yeast).

Gather these items to start:

- 4 cups of milk (cow, goat, or coconut milk)
- 2 tablespoons of kefir grains

Then follow these steps:

1. Pour the milk into a clean glass jar.
2. Add the kefir grains to the milk and stir gently.
3. Cover the jar with a cloth or paper towel and let it sit at room temperature for up to 24 hours.
4. Once the milk has thickened and turned tangy, strain out the kefir grains and bottle the kefir. Store it in the fridge.

Project: Create a Fermented Foods Recipe Book

Here's your project for this chapter: starting your own Fermented Foods Recipe Book. This is where you'll document the recipes you try, your modifications, and any tips and tricks you learn along the way.

Here's what you'll need to do:

1. Divide your recipe book into sections based on the type of fermentation (like vegetables, drinks, dairy, meats, etc.)
2. For each recipe, write down all the ingredients, measurements, and step-by-step instructions.
3. Make note of how long each recipe takes to ferment and how it changes over time.
4. Keep track of the results, like taste, texture, and any problems that arise. This way, you can repeat your successes and learn from your mistakes.

Key Points To Remember:

- Fermentation is an ancient method of preserving food that involves breaking down sugars with microbes like bacteria, yeasts, or molds.
- Fermented foods are packed with probiotics that improve digestion, boost your immune system, and even support mental health.

- When you ferment food, you unlock more nutrients and reduce substances that might be hard for your body to digest.
- Fermentation helps preserve foods like cabbage and cucumbers, which turns them into nutrient-dense and long-lasting treats.
- Lactic acid fermentation (think yogurt and sauerkraut) creates tangy flavors while keeping harmful bacteria away, making it great for long-term storage.
- Alcoholic fermentation (used in beer, wine, and bread) breaks down sugars into alcohol and carbon dioxide but is less common for long-term food storage.

BOOK 18:

HOMESTEADING AND SUSTAINABLE LIVING

Chapter 1: Designing Off-Grid Homes

When it comes to prepping for disasters, one of the most critical things to consider is your shelter. That's because a traditional home might not always be an option in the aftermath of a major event, especially if you are forced to bug out (which you might need to do in the event of a natural disaster like a hurricane or an earthquake).

How to Build an Off-Grid Shelter Using Natural Materials

When you're living off the grid, you need a shelter that's durable, efficient, and doesn't rely on modern infrastructure like electricity or plumbing. The beauty of using natural materials is that they not only offer sustainability but also blend in with the environment, ensuring your home is part of the natural landscape.

Examples of materials you can use to build your shelter include:

- Wood
    - Wood is one of the most versatile materials for off-grid homes. Look into traditional timber framing techniques like log cabins or post-and-beam construction for strength and insulation.
- Stone
    - Stone is durable, weather-resistant, and incredibly energy-efficient. It's perfect for building walls, foundations, and even chimneys.
- Straw Bale
    - Straw bale construction is becoming increasingly popular in off-grid homes. That's because straw is a renewable resource, offers great insulation, and is surprisingly fire-resistant.

- Cob
  - Cob is a mixture of clay, sand, and straw, which can be molded into walls. It's durable and provides excellent thermal mass, which means it absorbs and slowly releases heat throughout the day and night.
- Adobe
  - Like cob, adobe is made from earth (typically clay, silt, and sand) and is shaped into bricks for construction. Adobe homes are well-suited to warm climates.
- Earthbag
  - Earthbag homes are built by filling polypropylene bags with soil, and then stacking them like bricks to form walls. These homes are surprisingly sturdy and have great insulating properties. Earthbag construction is also relatively easy to do by hand, making it an excellent option for those looking to build their own shelter.

Project: Plan and Design Your Off-Grid Home

Now that you've got the basics down, it's time to start planning your off-grid home! Here's a step-by-step guide to help you design your own shelter.

1. Start by thinking about the size, purpose, and location of your off-grid home. Are you looking to build a small cabin for weekend getaways or a full-time home? Do you want it to be mobile (like a tiny house) or permanent? What kind of climate are you building in?

2. Based on your location and the materials available to you, decide what natural materials you want to use for the build. Think about how each material will perform in your climate. Will it keep you warm in the winter? Cool in the summer? How much maintenance will it require?

3. Draw a basic layout of your home, and consider space for living, sleeping, cooking, and storage. Don't forget to plan for things like water collection, waste management, and solar power.

4. Think about how your off-grid home will minimize its environmental impact. Are there renewable energy options like solar or wind power that you can incorporate? Can you design the home to take advantage of natural heating and cooling? How will you handle water and waste?

5. Once you've got your basic ideas down, create a sketch of your off-grid home. Don't worry about getting it perfect—just focus on getting your ideas on paper so you can start working out the details.

Chapter 2:
Waste Management and Recycling

When you're living off the grid, managing waste is one of the most important)and often overlooked) considerations. In a traditional home, flushing a toilet or tossing waste in a trash can is simple and convenient, but when the grid goes down (whether due to an EMP attack, a hurricane, or a long-term power outage), you need to have sustainable systems in place to handle waste and resources.

Composting Toilets and Greywater Systems

One of the cornerstones of off-grid waste management is a composting toilet. These toilets allow you to handle human waste without the need for a traditional plumbing system.

Composting Toilets

Composting toilets come in many designs, but the basics are simple. You have a toilet seat over a composting bin, and the waste is mixed with materials like sawdust or coconut coir to promote decomposition. Over time, the waste breaks down into nutrient-rich compost that can be safely disposed of or used to fertilize non-edible plants.

When my family lived through a power outage after a massive snowstorm, we were lucky to have a composting toilet installed. While our neighbors were dealing with frozen pipes and septic issues, we were still able to use our toilet without relying on the grid. The lack of water for flushing wasn't an issue, because with just a scoop of sawdust we were good to go.

Greywater Systems

A greywater system collects wastewater from sinks, showers, and washing machines (but not from toilets or kitchen sinks) and uses it for irrigation or other non-potable purposes.

You can design a simple greywater system by running the water from your shower or sink through a filter and then diverting it to your garden or composting toilet. This system is especially helpful in off-grid living situations.

I've set up a basic greywater system at my place, using an old bathtub and some PVC pipe to direct the water to my garden. Not only did it keep the plants happy, but it also saved us a ton of water.

Managing Sewage and Waste Disposal Off-Grid

When living off-grid, you don't have access to the usual sewage treatment plants. That means you need to be creative in how you manage waste. Here are some options:

- Septic Systems
  - If you have the space and resources, a small septic system can work well in rural, off-grid locations. However, they require proper maintenance and are dependent on having the right kind of soil for filtration. I once had a failed septic system during a particularly heavy rain, and it was a huge headache. After that, I made sure we had a backup composting toilet system in place for emergencies.
- Biogas Toilets
  - If you want to take it a step further, biogas toilets can help process waste into biogas, which you can use for cooking or heating.
- Pit Latrines

  - For more remote locations or emergency situations (like following an EMP attack where utilities are down), a simple pit latrine can suffice. It's essential to dig the pit in a location far from any water sources to prevent contamination.

- Wastewater Treatment Systems
  - A more advanced option is a small-scale wastewater treatment system. These can be designed to filter and clean wastewater from sinks and showers. They're usually powered by solar energy and use natural filtration processes like sand and gravel.

Project: Build a Simple Composting Toilet

Ready to dive in? Let's build a simple composting toilet you can use in an off-grid situation.

Here's what you'll need:

- A toilet seat
- A large plastic bin or bucket (around 5 to 6 gallons in size).
- Sawdust, coconut coir, or peat moss. You'll need about 2 to 3 gallons of material.
- A small ventilation pipe (optional, but helps with odor control).
- To line the container for easy waste disposal.
- For adding composting material after each use.

Then follow these steps:

1. Clean and sanitize the container, and then line it with a compostable liner or trash bag.
2. Secure the toilet seat to the top of the container. You can use screws or even just a sturdy lid to hold it in place.
3. Place your composting material (like sawdust or coconut coir) next to the toilet for easy access. The idea here is to add a scoop of material after each use to cover the waste and help it decompose.
4. If you're planning on using the toilet long-term, you can add a small ventilation pipe at the top of the container to help with odor control.

Source

After those steps, once the container is full, you can dispose of the waste in a compost pile (away from food gardens) or in a dedicated composting system.

Chapter 3:
Sustainable Living for Long-Term Survival

Living sustainably isn't just about minimizing your ecological footprint. It's about ensuring your very survival, and especially when you're off the grid. I'll share some of my own experiences with sustainability, like the time I had to rely on solar power after a freak snowstorm knocked out the grid, and how I managed to keep my family comfortable without relying on the usual utilities.

Reducing Your Ecological Footprint

The first step in sustainable living is reducing your ecological footprint...basically, the amount of natural resources you use and the waste you produce. Every decision you make, from how you source your food to how you dispose of waste, has an impact on the environment. The goal here is to minimize that impact and live in a way that doesn't deplete the planet's resources.

Waste Reduction:

- Recycling and Repurposing
  - One of the easiest ways to reduce your ecological footprint is by recycling and repurposing materials. Metal, glass, paper, and plastic can be recycled or reused for various purposes. I've turned old glass jars into storage containers and used scrap wood to build raised garden beds.
- Composting
  - By composting organic waste, you return valuable nutrients to the soil and reduce the amount of garbage that goes into landfills.
- Zero-Waste Lifestyle
  - You can also embrace a zero-waste mindset by reducing packaging, using reusable containers, and opting for sustainable products like bamboo toothbrushes and cloth

shopping bags. You'd be amazed at how quickly you can cut down on single-use plastics with just a few changes in daily habits.

During one of the power outages in my area, I realized how much unnecessary waste I had been accumulating without even thinking about it. We switched to using cloth napkins, stopped using disposable water bottles, and even started making our own cleaning products.

Techniques for Long-Term Self-Sufficiency

Here are a few tips I have for ensuring that you can be self-sufficient over the long term:

Food Production

- Growing your own food is one of the most reliable ways to achieve self-sufficiency. Start by growing easy-to-care-for crops like tomatoes, beans, and root vegetables.
- Permaculture is a holistic approach to farming that focuses on sustainability and working with nature. By using companion planting, mulching, and water-saving techniques, you can build a resilient food system that requires minimal intervention on your part.
- If you have the space and resources, raising small animals like chickens or goats can provide you with eggs, milk, and meat. Goats are especially useful for off-grid living because they can graze on the land and are relatively low-maintenance.

Water Collection and Purification

- Setting up a rainwater collection system using gutters, barrels, and filtration systems can provide you with a reliable water source. Once you've filtered the water, it's ready for drinking, cooking, or irrigation.

- Even if you have access to natural water sources, you'll need a filtration system to ensure it's safe to drink. Solar-powered UV water purifiers or gravity-fed filters are great options for off-grid living.

Waste Management

- As I mentioned in the previous chapter, composting is a great way to manage organic waste while creating valuable compost for your garden. Composting toilets are also an excellent solution for off-grid waste management.

- A biogas digester can turn organic waste into methane gas, which can be used for cooking or heating. Setting up a small biogas system is a great way to turn waste into a resource and further reduce your reliance on outside energy sources.

Project: Set Up a Solar-Powered Water Heating System

One of the most useful and sustainable systems you can set up in your off-grid home is a solar-powered water heating system. Here's a simple guide to get you started:

Here's what you'll need:

- Solar Collector
    - This is the part of the system that captures the sunlight and converts it into heat. You can buy pre-made collectors, or you can build one yourself using copper tubing and black-painted metal or plastic for optimal heat absorption.

- Water Tank
    - A large insulated tank (preferably made of durable material) will store the hot water. This tank should be placed near the solar collector to ensure efficient heat transfer.

- Pipes and Pump
    - Pipes will connect the solar collector to the tank. A small and energy-efficient pump will help circulate the water.

- Thermostat or Temperature Controller
  - This will help you monitor the water temperature and control the pump to prevent overheating.

Then follow these steps:

1. To start, mount the solar collector on the roof or in a location that receives full sun for most of the day. Make sure it faces south (in the Northern Hemisphere) for optimal sunlight exposure.
2. Place your water storage tank near the solar collector, ideally in a shaded area to prevent it from losing heat.
3. Use pipes to connect the solar collector to the tank. Install the pump to help circulate the water.
4. Use a thermostat or temperature controller to regulate the water temperature and prevent overheating.
5. Once everything is installed, test the system to ensure it's working properly. You should see hot water circulating from the collector to the tank.

Key Points To Remember:

- Off-grid homes built with natural materials like wood, stone, straw bales, and cob not only blend with the environment but also reduce your reliance on modern infrastructure.
- Waste management is crucial for off-grid living, so systems like composting toilets and greywater setups are essential for sustainability and hygiene.
- Creating self-sufficiency through food production is key—grow your own crops, raise animals, and embrace permaculture techniques to ensure long-term survival.
- Solar power and other renewable energy sources are the backbone of off-grid living, providing power for everything from water heating to cooking.
- Reduce your ecological footprint by recycling, composting, and embracing a zero-waste lifestyle, ensuring you use resources wisely and sustainably.
- Having backup systems for waste disposal, like biogas toilets or septic systems, can prevent headaches when living off the grid.

BOOK 19: REBUILDING COMMUNITIES POST-DISASTER

Chapter 1:
Organizing Community Resources

Your community's resilience can make or break your survival. Having a network of neighbors, friends, and local groups who can pool resources, share skills, and support each other.

Establishing Community Resource Centers

A community resource center acts as a hub where people can access essential supplies, information, and support. In times of disaster, it's important to have a central place where people can go for help, whether that's food, medicine, or tools.

Start by identifying a location that can serve as a resource center; this could be a community hall, church, or even a local school. The goal is to create a space where people can store essential supplies like food, water, first aid kits, and tools.

I've seen firsthand how valuable a community center can be. During a flood in my area a few years ago, a local church opened its doors as a resource center. They distributed food, water, and blankets, and the community came together to help each other. It was a real eye-opener and reinforced the idea to me that having a well-organized community can really save or improve lives when things go wrong.

Building a Local Supply Chain for Sustainable Living

Building a local supply chain is essential for ensuring your community's long-term survival, especially when outside help might not be available for an extended period.

I would work with local farmers and gardeners to establish food-growing networks. Encourage urban gardening, small-scale farming, or even communal farms. For basic needs like clothing, household items, or construction materials, you can also connect with local makers or artisans who can produce goods within the community.

I've seen local supply chains work wonders during tough times. After a hurricane, for instance, local farmers teamed up with nearby community groups to create a barter system where people traded food and goods.

Chapter 2: Establishing Trade and Barter Networks

When the grid goes down, the traditional monetary system may no longer function. That's where a local barter system becomes invaluable.

Establishing Trade Guidelines and Fair Exchanges

Once you and the other people involved have established the barter system, you'll need to create trade guidelines that ensure fair exchanges. Without currency, it's easy for unfair trades to happen.

- Determine how you'll value goods and services. One way to do this is by establishing a unit of exchange, such as "1 can of beans equals 1 jar of honey." While you might not always have an exact 1:1 ratio, having rough guidelines will help ensure everyone feels like they're getting a fair deal.
- Make sure that people are transparent about the quality and condition of the items they're offering.
- If you have a skill or service to offer, let others know that you're willing to trade your labor for essential goods. This can include everything from fixing plumbing, to teaching classes on food preservation, to offering medical care.

I remember during a major hurricane, my neighbors and I created a "barter board" to post what we needed and what we had available. People traded everything from generators to jars of homemade jam.

The Role of Community Currency After Collapse

In a complete collapse of the financial system, money might no longer hold value. That's where community currency can play a role. A community currency is like a local form of money that can be used for trade and exchange within the community. It helps streamline bartering by assigning value to goods and services.

Community currencies can take many forms, from physical tokens, to paper notes, to digital credits. Ideally, a community currency should be backed by actual goods or services, such as food, energy, or labor. This ensures that the currency holds real value and can be used for trade when needed.

For a community currency to work, everyone in the community must agree to use it and accept it for trade. This is why it's imperative to establish the system early and get the community on board.

Unlike traditional currency, a community currency should be controlled to prevent inflation. Since it's usually tied to the local economy, the value of the currency will fluctuate based on supply and demand, so it's important to manage it wisely.

During an extended power outage, I witnessed how some communities began using barter tokens to keep things organized. For example, we traded water and food using tokens, which acted as a unit of value for the goods people were offering. It made things smoother and more predictable.

Project: Set Up a Local Barter Market

A local barter market is a great way to get your community engaged and start practicing the barter system in a controlled environment.

Here's how you can get started

1. Pick a central location that's easily accessible for everyone. A community hall, church parking lot, or public park can work well for a market-style setup. Make sure it's large enough to accommodate booths and tables.

2. Set up designated areas for different types of goods like food, tools, clothing, and services. If you're incorporating skills trading, you can designate an area where people can exchange knowledge (like cooking classes or first aid training).

3. Make sure everyone knows the rules for trading. Do you want to use community currency, or will trades be based on direct exchange? Establish a fair trade system, and ensure everyone understands it. Provide a simple brochure or flyer outlining the guidelines.

4. Spread the word through flyers, word-of-mouth, and local social media platforms (if available).

5. Hold regular barter market days. The first few events will be a learning experience, but they'll build momentum and get people excited about participating.

Chapter 3:
Community Health and Education

When the world as we know it crumbles one of the first things you'll want to focus on is your community's health and education.

After all, survival isn't just about food, water, and shelter. Even though those things are important, but also about ensuring that future generations can live on and thrive ... .even when everything else feels like it's falling apart.

Trust me, this is more important than you might think. When the internet's down, and the local school is closed, the only way to ensure your community thrives is to educate and care for each other the old-fashioned way.

Teaching Survival Skills to Future Generations

The key to long-term survival, especially in an off-grid world, is making sure that your kids (and the kids of your community) grow up knowing how to thrive when the systems we rely on today no longer exist.

Sure, it's tempting to let them binge-watch the latest streaming show (we've all been there), but in an emergency, knowing how to start a fire or identify edible plants will be far more valuable.

The more skills kids learn, the less they'll depend on others. They'll be able to grow their own food, build shelter, and take care of themselves and others.

Teaching survival skills fosters critical thinking and creativity. When faced with a problem, kids with these skills can think on their feet and come up with practical solutions.

There's something deeply grounding about learning how to survive in the wild. The more kids are connected to the land, the more they'll respect it.

Here are the most important skills that I would teach:

- Fire Starting
    - This could be as basic as teaching them how to use matches or a flint striker, or as advanced as building a fire without tools. Make it fun by challenging them to start a fire with just sticks and friction. Think of it as a life-or-death game of "Survivor."
- Foraging and Hunting
    - Learn to identify wild edibles (wild greens, berries, mushrooms) and teach your kids the same. Not only will they be able to find food when supplies are low, but it'll also give them a deep sense of accomplishment. Plus, foraging is an excuse to get out and explore the outdoors, and who doesn't like a nature hike?
- Water Purification
    - Teach the kiddos how to filter water using basic materials (like rocks, sand, and charcoal) and then boil it to make it safe to drink. They'll feel like little mad scientists, and you'll be confident they can survive even if the water system goes down.
- Basic First Aid
    - I'm talking about the basics: how to clean a wound, apply a bandage, treat burns, and recognize the signs of hypothermia or dehydration. These skills can save lives.

I remember teaching my son how to build a fire during a camping trip gone awry. We got caught in a freak thunderstorm, and our tent got soaked. My son, about 10 at the time, was *adamant* that we could start a fire, and sure enough, with a little patience, some twigs, and a flint, we had a warm blaze going. That was a huge moment for me; he was learning something that could save his life one day, and we were doing it together.

Managing Community Health and Sanitation

A healthy community is a resilient community, and that starts with managing health and sanitation like we covered previously in this book

Here are some basic sanitation practices that I would make sure your team is

- Composting Toilets
  - Like I covered in an earlier chapter, composting toilets are amazing when traditional sewage systems are unavailable.
- Clean Water
  - Always ensure that water is purified before consumption. Setting up a filtration system, along with knowing how to boil water, should be taught to everyone. Even having some water purification tablets on hand can be a lifesaver.
- Waste Management
  - In the absence of modern waste collection, it's a good idea to separate recyclables, compost, and non-recyclable waste to keep things organized and hygienic.

Project: Create a Community Health and Education Program

Now it's time to put everything you've learned into action with a hands-on project: creating your own Community Health and Education Program.

Here's what you'll need to do:

1. What skills and knowledge does your community need most? Start by gathering a group of community members and discussing what areas need improvement. Is it health education, survival skills, farming, or something else?
2. Based on the needs identified, organize regular workshops or classes to teach essential skills. These could include first aid, gardening, food preservation, or home maintenance. Be creative! The more interactive, the better.
3. With limited resources, use what you have available to make teaching materials. You can create charts, diagrams, or simple handouts to illustrate important points. Get everyone involved in making these materials!

4. Establish a community health program that's focused on hygiene, waste management, and disease prevention. This could involve setting up a hand-washing station, educating the community on food safety, and organizing local health check-ups

Key Points To Remember:

1. A community resource center is a game-changer when disaster strikes because it will be your go-to spot for supplies, info, and support to keep everyone afloat.
2. Building a local supply chain with farmers and artisans helps your community stay self-sufficient, so you're not waiting for outside help that might never come.
3. Bartering and community currency are going to be your best friends when cash is worthless, since trading goods and services keeps the wheels turning.
4. Hosting a local barter market is a great way to get people involved in trading, set some fair rules, and help everyone get what they need without the stress.
5. Teaching kids basic survival skills (like making fire, finding food, and purifying water) isn't just fun, it's essential for making sure they can handle anything that comes their way.

BOOK 20: MAKING YOUR OWN DEFENSE TOOLS

Chapter 1:
Crafting Survival Weapons

When the world goes sideways one of your top priorities is going to be protection and hunting.

Making Weapons

When it comes to survival, a good knife is a must-have. But what if you don't have access to a store-bought one? Well, crafting your own could be one of the most valuable skills you can learn.

Crafting a Survival Knife

You don't need fancy equipment to make a functional knife. For a simple knife, you can use high-carbon steel (old saw blades, files, or even scrap metal). Then follow these steps:

- Shaping the Blade
    - Start by cutting your steel into the shape of a knife blade. If you're using a file or old saw blade, make sure to wear protective gear because those materials can be tough to handle.
- Crafting the Handle
    - For the handle, you can use wood, bone, or even antler. Shaping the handle to fit comfortably in your hand is key, so don't rush through this part.
- Sharpening
    - Once you've got the shape, it's time to sharpen the blade. A sharpening stone or even the side of a rock will do in a pinch. You want a fine edge for cutting, so take your time and make sure it's sharp enough to do serious work.

I've made a few knives over the years using scrap metal from old junk around the farm. The first one I crafted was a bit rough around the edges (pun intended), but it got the job done when I needed it most. Plus, there's a kind of pride that comes with making your own tools from scratch.

Crafting a Spear

A spear is another versatile weapon that's easy to make and can serve both as a tool for hunting and as a defensive weapon. A simple spear consists of a long and sturdy stick as well as a sharpened point (either made of wood or metal).

- Select the Right Wood
    - Choose a strong and straight piece of wood; ideally hardwood like oak, ash, or hickory. The shaft should be around 5 to 7 feet long.
- Sharpen the Tip
    - For the spearhead, you can carve a sharp point into the wood. Alternatively, you could attach a stone, bone, or metal tip for added durability.
- Strengthen the Connection
    - If you're attaching a metal or stone head, use string, sinew, or strong twine to secure the tip to the shaft. You'll want a firm hold so that it doesn't come loose when you're using it.

Crafting a Slingshot

A slingshot is an effective weapon for both defense and small game hunting. You can craft one using wood for the frame, rubber bands or inner tubes for the elastic, and a piece of leather or fabric for the pouch to hold the projectiles (like small rocks or lead ammo).

- Find a Suitable Branch
    - The frame of your slingshot needs to be Y-shaped, so find a branch that's strong and has the right shape. A hardwood branch works best I've found.
- Add the Elastic
    - You can use rubber bands, or if you're lucky, find an old inner tube from a bicycle or car tire. Cut strips of the rubber and attach them to each side of the Y-frame.
- Make the Pouch

- Cut a piece of leather or sturdy fabric and attach it to the rubber strips. This is where you'll place your projectiles (like small rocks or marbles).

Project: Build Your Own Survival Knife

Now that you've got the basics down, it's time for a hands-on project. Building your own survival knife is a useful and rewarding task. Here's a step-by-step guide to help you craft a simple survival knife that can serve you well in any situation.

Here's what you'll need:

- High-carbon steel (or scrap metal from an old saw blade or file)
- Wood, bone, or antler (for the handle)
- A file or grinder (to shape the blade)
- Leather or cord (for wrapping the handle)

Then you can follow these steps:

1. Start by cutting your steel into a rough shape of your desired knife. You can use a file or grinder to shape it, but make sure the edge is sharp and free of nicks.
2. Use a sharpening stone to refine the blade. Keep working on it until you have a smooth and sharp edge that can cut through wood, rope, or anything else you need it to.
3. Choose a material for the handle; wood, bone, or antler work well. Shape the handle to fit comfortably in your hand and attach it to the blade using strong adhesive or by wrapping it tightly with leather or cord.

Chapter 2:
Building Traps and Defensive Devices

In a survival situation, setting up traps and defensive devices is critical for both food acquisition and protection.

Creating Simple Traps for Animals and Intruders

Deadfall Trap

This is one of the easiest traps to make. It uses a heavy object (like a rock or log) that falls when triggered.

Follow these steps:

1. Find a flat surface and place a heavy rock or log above your bait.
2. Prop the rock up with a stick (the trigger), ensuring it's stable but will fall when disturbed.
3. Place bait (such as nuts or a piece of meat) under the rock, and when the animal tries to grab it, the trap will trigger.

Box Trap

The box trap is simple but effective for catching animals like rabbits, raccoons, or squirrels.

Follow these steps:

1. A box or crate large enough for your target animal to fit in.
2. A stick to prop up the box, with string attached to the stick and bait inside.
3. Once the animal enters to get the bait, the string releases the stick, causing the box to fall and trap the animal.

Cage Trap

A cage trap allows you to capture animals alive, which is useful for larger animals or when you need to release them later.

Follow these steps:

1. Shape a cage, ensuring the bars are close enough to trap the animal but far enough apart to allow entry.
2. Attach a spring-loaded door or a pressure plate that drops the door once the animal enters.
3. Place food or other bait inside to lure the animal into the cage.

Constructing Tripwires and Alarm Systems

Tripwires

Tripwires are ideal for setting up immediate alerts. You don't need complex tools; just wire, rope, or string.

Here's how to make one:

1. Stretch a thin and strong wire across a path, entrance, or area where you suspect intruders or animals will pass.
2. Connect the wire to an object (like a bell, tin can, or a spring-loaded mechanism) that will make a noise when triggered.
3. When someone or something crosses the wire, the object will fall or make noise, which alerts you to the presence of a threat.

Alarm Systems

Alarm systems can alert you to any unwanted presence, whether animal or human. The basic principle is to create a loud, disruptive noise when something crosses a sensor (such as a tripwire). To make a basic alarm system:

1. Use a tripwire attached to a tin can filled with rocks, or a bell.
2. Tie the tripwire to the object and stretch it across a path, doorway, or entryway.
3. When the wire is triggered, the can falls or the bell rings, letting you know there's movement.

Project: Set Up a DIY Defensive Trap

Let's take it one step further and build a defensive trap that's designed to alert you to movement and, if needed, disable or deter an intruder.

Here's what you'll need:

- Strong twine or fishing line
- A tin can or metal object (to create noise)
- A stick or branch (for the trigger mechanism)
- Rocks or heavy objects (to weight the system)
- A sturdy tree or pole (for support)

Then follow these steps:

1. Stretch the twine or fishing line across the entrance or path where you expect movement. Attach one end to a sturdy tree or pole, ensuring it's taut enough to trigger with light pressure.
2. Use a stick or branch as the trigger. Place it in such a way that if the wire is pulled, the stick falls and releases the metal object (your noise-making device).
3. Tie a tin can, metal container, or bell to the other end of the tripwire. The idea is that when the tripwire is triggered, the noise device will fall, clanging loudly to alert you of movement.
4. To make the trap more efficient, you can attach a weight or a heavy object to the tripwire to give it more force when pulled. This can help pull the metal object down harder or even set off a secondary device if desired.

Chapter 3:
Improvised Weapons for Self-Defense

When things go sideways, you might not have access to the typical weapons you'd rely on. That's when improvising with everyday objects becomes not just practical but necessary for self-defense.

Using Everyday Objects as Weapons

In a survival situation, your environment is full of potential weapons. You don't need to look far for something that can protect you, and regardless of whether you're defending yourself or deterring an intruder.

Common items you can use as weapons include:

- Flashlights
  - A heavy-duty flashlight can serve as a blunt object for striking, or you can use it as a makeshift torch to light your way (or intimidate).
- Pens
  - A sturdy pen can be used as an improvised stabbing weapon. If you're ever caught in a situation where you need to defend yourself, a pen can become a surprising and effective weapon.
- Rocks and Bricks
  - Found anywhere, rocks and bricks can be thrown as projectiles or used for defense if you're under attack.
- Belt
  - A belt with a heavy buckle can be swung like a club, or used to choke an attacker if necessary.

- Keys
    - Held between your fingers, keys make a makeshift stabbing weapon, which is perfect for close-quarter defense.

Project: Craft a Blowgun and Ammo

Ready to make your own survival tool? Let's walk through how to craft a DIY blowgun and its ammo.

Here's what you'll need:

- A 3-4 ft PVC pipe
- Rubber tubing (for the mouthpiece)
- Sharp needles or nails (for ammo)
- Tape or glue
- Scissors or a knife (to cut the pipe)

Then follow these steps:

1. Measure and cut the PVC pipe to about 3-4 feet. Ensure the pipe is smooth inside for maximum airflow.
2. Take the rubber tubing or soft straw and attach it to one end of the pipe. You can secure it with tape or glue to ensure it doesn't slip off.
3. Sharpen wooden dowels, nails, or use sewing needles for your ammo. They should fit snugly inside the blowgun but also be light enough to fly accurately.
4. Once everything is assembled, test it out! Use a piece of paper or soft target to check the force of the darts. Adjust the ammo length or pipe length if needed for better accuracy.

Key Points To Remember:

1. Crafting your own survival weapons is a vital skill when you're off the grid or in a post-disaster world, and it starts with basic tools like knives, spears, and slingshots.
2. Making a survival knife is simple: all you need is high-carbon steel (think scrap metal), a good handle material like wood or bone, and patience to sharpen it to perfection.

3. Spears are incredibly versatile for both hunting and defense, just pick a strong piece of wood, sharpen the tip, and secure it with twine or sinew for durability.
4. Slingshots can be easily crafted from a Y-shaped branch, rubber bands, and leather to give you a compact and effective tool for small game hunting or defense.
5. Setting up traps like deadfalls, box traps, and tripwires helps with food acquisition and security. Basic, easy-to-make solutions for keeping threats at bay.
6. Improvised weapons are often at hand when you need them most. Think flashlights, pens, rocks, and belts, which can all be turned into effective tools for self-defense in a pinch.

BOOK 21: ADAPTING TO EXTREME WEATHER

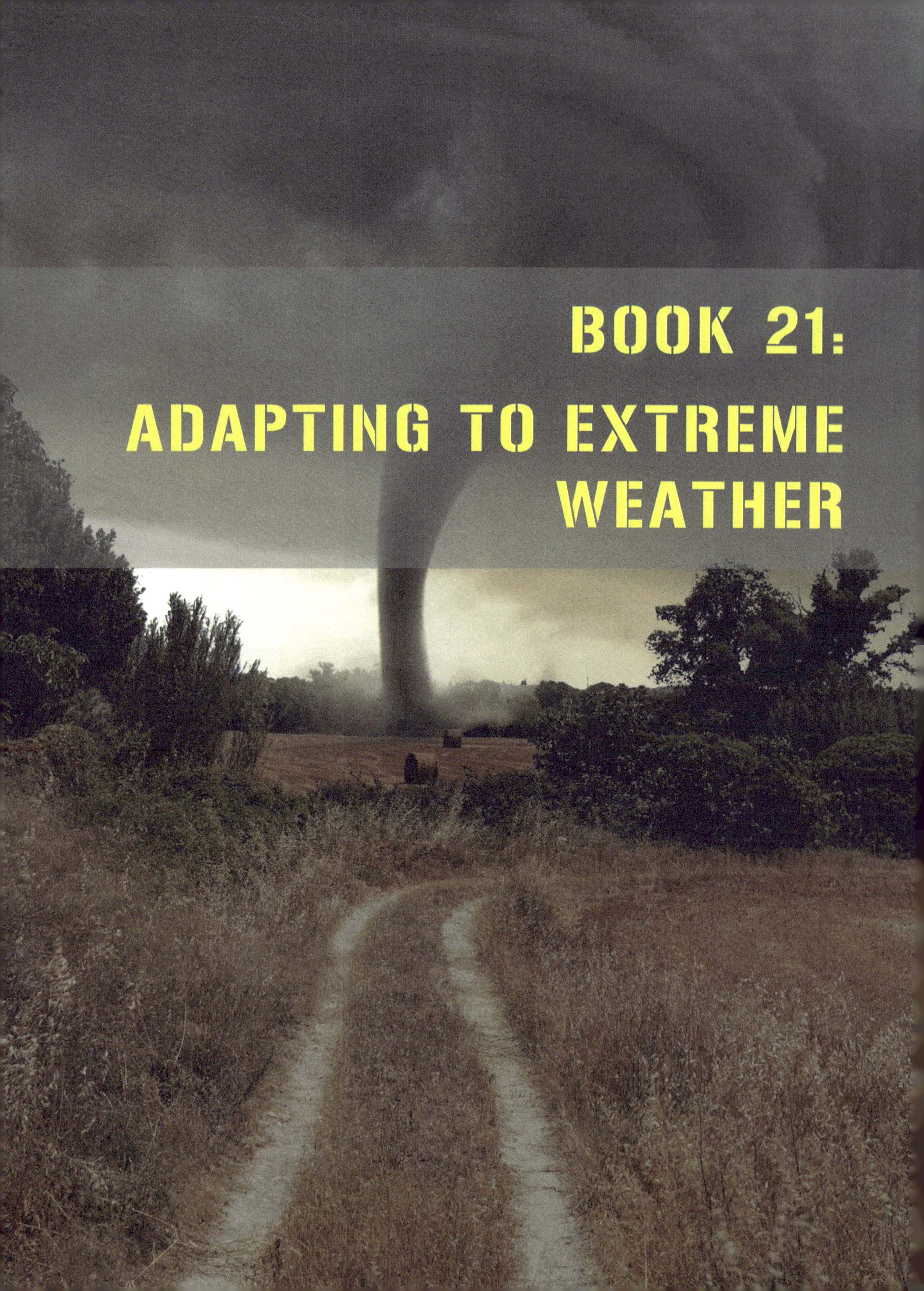

Chapter 1: Surviving Heatwaves and Drought

Heatwaves and droughts can be some of the most dangerous disasters to endure. Without proper planning, the extreme heat can lead to dehydration, exhaustion, and even heatstroke.

Techniques for Staying Cool and Hydrated in Extreme Heat

Staying cool in extreme heat is super important to avoid heat exhaustion or heatstroke. These conditions can sneak up on you, and the last thing you want is to be caught unprepared.

Here's what you can do:

- Wear Light, Loose Clothing
  - Light-colored and loose-fitting clothing helps your body breathe and keeps heat from building up around your skin. Avoid dark clothes as they absorb heat.
- Use Wet Cloths
  - One of the quickest ways to cool down is by using a wet cloth or bandana. Wet it with cold water and drape it over your neck, forehead, or wrists. These areas have major blood vessels close to the skin and can help cool you down quickly.
- Find Shade
  - Always seek shade or create your own if none is available. If you're out in the open, use a tarp or reflective materials to create a quick shade shelter. Being out of direct sunlight can drop the heat index by several degrees.
- Hydrate, Hydrate, Hydrate
  - Water is your best friend in extreme heat. Drinking small sips of water consistently is far more effective than gulping large amounts at once.

A few years ago, I went camping during a heatwave. My kids were having a blast, but I quickly realized we weren't prepared for the heat. We ended up soaking towels in cold water and laying under trees for shade. The lesson I learned that day? Always have a backup hydration and cooling plan.

Water Conservation and Heat Protection Strategies

When dealing with heatwaves and droughts, conserving water is just as critical as staying cool. You may not have an unlimited supply, so knowing how to use it efficiently can make a huge difference.

- Limit water use to essentials. Focus on drinking water and keeping cool, and minimize activities like washing dishes, clothes, or bathing. If you're in a family setting, designate certain times of day for essential water uses, and
- If you're outdoors, consider capturing rainwater or using greywater (water used in sinks or showers) for non-drinking purposes. Even a small rain barrel can catch enough water for watering plants or washing hands. Keep a few buckets around your home or campsite for collecting water from condensation, pools, or even condensation from coolers.

Last summer, we had a stretch of triple-digit days where our water supply started running low. We ended up using the kiddie pool water to rinse off, and I learned that even a small pool can be repurposed in a pinch! By thinking ahead and adjusting how we used our resources, we were able to stay safe and stretch our water until the supply came back online.

Chapter 2:
Surviving Cold Conditions

When the temperature plummets, survival isn't just about staying warm. It's about keeping your core temperature stable enough to stay alive. But don't panic, because I'll break down how you can build warm shelters, use insulation and thermal clothing, and make your own sleeping bag to keep yourself alive and comfortable in freezing temperatures.

Using Insulation and Thermal Clothing Effectively

Layering up isn't just a fashion statement. It's life or death in freezing conditions. Proper insulation traps your body heat and keeps the cold at bay. To make the most of your layers, follow these steps:

- Base Layer
    - This layer touches your skin, so it needs to wick moisture away from your body. Wool or synthetic materials like polyester work best. Avoid cotton, as it absorbs sweat and can make you colder.
- Middle Layer
    - Your mid-layer traps the heat. Fleece, down, or wool jackets are ideal for this. Remember, the goal is to trap warm air close to your body.
- Outer Layer
    - This is your shield from wind, rain, or snow. Waterproof and windproof jackets or ponchos are a must. A good outer layer keeps moisture and wind from robbing you of your warmth.
- Gloves and Hats

  - A lot of heat escapes from your head and hands. Don't skimp on gloves and hats made of insulated materials. In a pinch, you can always use a scarf or bandana to wrap your head.

- Boots and Socks
  - Insulated and waterproof boots are good, but don't forget socks! Wool socks are warm, even when wet, and layering thin socks under thick ones can prevent frostbite.

I had one of those "aha" moments while trying to survive a power outage during an unexpected early winter storm. We were without heat for several days, and I wasn't prepared with proper thermal gear. I ended up wearing every sweater I could find, but it wasn't until I layered up in wool socks, a down vest, and a good waterproof jacket that I stopped shivering. It was a lesson in layering—and also in humility. Never underestimate winter's ability to turn your world upside down.

Project: Make Your Own Insulated Sleeping Bag

Alright, let's get hands-on. A good sleeping bag is a lifesaver when it comes to staying warm in freezing conditions, but if you don't have one, you'll need to improvise.

Here's what you'll need:

- A large tarp or thick cloth (for the outer layer)

- Insulating materials like leaves, moss, or shredded paper (or if you have it, extra wool or fleece)
- A needle and strong thread or twine
- A sleeping mat or some form of soft padding (for the bottom layer)

Then follow these steps:

1. Lay your tarp or thick cloth flat. This will be your sleeping bag's outer shell, and it needs to be long enough to wrap around you.
2. Start collecting your insulation. If you're in a forest, gather leaves or moss. You can also use extra clothing or fabric to create padding. Pack these materials inside the cloth, leaving room for air to circulate for maximum insulation.
3. Fold the cloth and sew or tie it into a bag shape, leaving an opening at the top. Make sure there's enough space to move around comfortably, but not too much that your body heat won't keep the insulation warm.
4. Tie the open ends with twine or fabric, but make sure the bag stays breathable. You can also add an extra layer around your feet for added warmth.
5. If you have a sleeping pad, place it inside for added insulation from the cold ground.

Source

Chapter 3: Surviving Storms and Tornadoes

When severe weather strikes, everything changes in an instant.

Preparing for and Surviving High Winds and Floods

High winds and floods are often the first signs of an impending storm. Depending on where you are, you might experience one or both at the same time.

High Winds

Whether it's a tornado, hurricane, or even just a powerful thunderstorm, high winds can quickly turn dangerous. The key is to secure your home and prepare yourself:

- Reinforce Doors and Windows: High winds can easily break windows or blow open doors. Use plywood to board up windows or install storm shutters to keep glass from shattering. Make sure your doors are tightly secured.
- Anything outside that isn't secured (lawn furniture, tools, trash cans) can become a projectile. Tie down or bring inside anything that could get tossed around by the wind. I learned this the hard way when a surprise storm picked up a plastic table and tossed it straight into our window. After that, I secured everything I could.
- If you're caught indoors during high winds, head to an interior room on the lowest level, away from windows. If you don't have a basement, a small, windowless bathroom or closet is your best bet. Stay low to the ground to avoid flying debris.

Floods

Flooding can happen fast, and it's often the result of a combination of heavy rain and rising water levels. When you live in an area prone to flooding, it's essential to know what to do ahead of time.

- Don't wait until the waters rise to evacuate. Have a plan for where you'll go, and leave as soon as authorities issue an evacuation order. Trust me, when my neighborhood flooded a few years ago, we waited too long to leave, and the roads quickly became impassable. The last-minute scramble was chaotic, and I've never made that mistake again.
- Elevate essential items, like electronics and important documents, to higher ground. If you can, move furniture away from flood-prone areas. If you're caught in a flood, you want to be as prepared as possible to keep valuable items safe.
- If evacuation isn't an option, have a flood emergency kit ready with essentials like food, water, medications, and a flashlight. Keep it in a waterproof bag, and keep the kit easily accessible.

Project: Install Storm Shelters in Your Home

Now that we've covered the basics of storm survival, let's talk about installing a storm shelter in your home, especially if you live in an area that's prone to tornadoes, hurricanes, or severe wind storms.

Here's what you'll need:

- Concrete or reinforced steel panels (for the shelter walls)
- Steel door with heavy-duty hinges (for the entrance)
- Plywood, blankets, or foam for insulation
- Basic tools: saw, hammer, drill, screws, etc.
- Water-resistant materials

Then follow these steps:

1. If possible, build the shelter in your basement or a central, windowless room. Make sure it's near your water and food storage so you're ready to hunker down for a few days.

2. Clear out the area where the shelter will go. If you're building a shelter in a basement, clear out space for the structure to fit. Reinforce the walls and foundation of the room using plywood or steel.

3. Use concrete or steel to create walls that will withstand high winds. Install a steel door with a secure locking mechanism to ensure safety.

4. Line the shelter with foam or blankets for insulation. Make sure to have enough food, water, and first-aid supplies for everyone who might use the shelter.

Key Points To Remember:

1. In extreme heat, staying cool and hydrated is critical, so wear light, loose clothing, seek shade, and drink small sips of water regularly to avoid heat exhaustion.

2. Water conservation is key in a heatwave or drought, so focus on essential uses like drinking and cooling off, and repurpose rainwater or greywater for other needs.

3. When the temperature drops, layering is essential for survival, so start with moisture-wicking base layers, add insulating middle layers, and finish with a waterproof outer layer to block wind and rain.

4. Make your own insulated sleeping bag using materials like leaves, moss, or extra wool to stay warm in freezing conditions without relying on store-bought gear.

5. For storms and tornadoes, reinforce windows and doors, secure outdoor items, and have a plan for evacuation or shelter in case of floods or high winds.

6. Install a storm shelter in your home, especially if you live in an area prone to tornadoes, hurricanes, or severe winds, and make sure it's stocked with essentials like food, water, and first aid.

BOOK 22: PSYCHOLOGICAL RESILIENCE IN A CRISIS

Chapter 1:
Mental Strength for Survival

Physical preparations are important, but what about your mind? You could have all the survival gear in the world, but if your mental resilience isn't solid, it's going to be a lot harder to make it through. Mental strength is what will keep you calm in the chaos and a lot more focused in moments of panic. That's what this chapter is all about.

Building Psychological Resilience for Adversity

Building mental strength doesn't happen overnight. It's a skill, much like learning to start a fire with nothing but friction or to craft a shelter from scratch. It's a muscle you train, and it's a key component of survival.

Stay Present

In the midst of chaos, the most important thing you can do is stay present. This means resisting the urge to panic, stay focused on what's happening in the moment, and tackle immediate needs one by one. I learned this lesson the hard way during a wildfire evacuation. As soon as I heard the news, my mind spiraled into "what ifs." But focusing on what I could control (like gathering supplies and securing my family) was the only thing that kept us grounded.

Mindset Matters

A survival mindset is about embracing the reality of the situation and accepting that things might be bad, but not letting that feeling take control. One way to build this mindset is by practicing positive self-talk. When I've been through tough situations, I'll remind myself: "This is tough, but I can handle it. I've got what it takes to get through this."

Resilience Through Adaptability

Resilience is also about being adaptable. Life rarely goes according to plan. So, when things go sideways (like that time my vehicle broke down miles away from help), it's the ability to adjust, think on your feet, and keep moving forward that will get you through.

Coping Mechanisms During a Crisis

When you're in the thick of it, a crisis can feel like an overwhelming tidal wave. But having a set of coping mechanisms can make all the difference.

Breathing Exercises

It sounds simple, but deep breathing can have an immediate calming effect when your heart rate spikes and stress sets in. Slow, deep breaths help lower cortisol levels (the stress hormone) and bring clarity. I remember being in the midst of a blizzard with no power and everyone on edge. A few deep breaths were just enough to keep me from losing my cool.

Break Tasks into Smaller Steps

When everything feels like too much, break tasks down into smaller, manageable steps. Don't think about the entire situation at once. Focus on the next right step. For example, during a major hurricane, I focused on securing windows, then gathering supplies, and then checking on my neighbors. Small wins lead to bigger successes.

Stay Connected

Isolation is one of the quickest ways to feel overwhelmed. During a long power outage after a storm, I realized how much I depended on regular communication, whether it was through a simple phone call or just checking in with my neighbors. Even when things go sideways, staying connected with others is crucial for keeping your morale high.

Maintain Routines

The crisis may throw everything out of whack, but maintaining a routine helps maintain a sense of normalcy. It doesn't matter if it's waking up, eating, or setting aside time to meditate or read; having a daily routine helps you retain some control.

When my family was cooped up during a snowstorm, we stuck to a routine: morning breakfast, followed by chores, and then evening family games. It kept everyone calm and grounded.

Chapter 2:
Psychological First Aid

When disaster strikes, not just the physical survival that matters. The emotional toll can be just as damaging. People will experience stress, anxiety, and trauma, and it's important to be prepared to help them cope. The same goes for you and your family.

Helping Others Cope with Stress and Anxiety

When people face extreme stress (whether it's from loss, general uncertainty, or the trauma of a disaster) it's easy for them to spiral into anxiety, panic, or even depression. But there are effective ways to help them navigate these tough emotions.

Active Listening

One of the best ways to help someone during a stressful time is to listen. It sounds simple, but truly listening without judgment can be a huge comfort. I've found that when someone is upset or anxious, just letting them express themselves without interruption often makes them feel heard and less alone. Don't try to fix everything immediately. Sometimes, all they need is someone to hear them out.

Provide Reassurance

After a disaster, people are often uncertain about what comes next. Providing reassurance is important. Tell them things like, "We're going to get through this together," or "It's okay to feel scared right now."

This helps normalize their feelings and takes the weight off of their shoulders. When we lost power for a week during a snowstorm a few years ago, my neighbors were all feeling panicked. A few reassuring words helped calm them down until we figured out our next steps.

Grounding Techniques

For people experiencing panic or overwhelming stress, grounding exercises can be extremely helpful. These techniques help someone focus on the present moment, moving them away from thoughts of impending danger or catastrophe.

A simple method is the 5-4-3-2-1 technique, where they name 5 things they can see, 4 things they can touch, 3 things they can hear, 2 things they can smell, and 1 thing they can taste. This engages the senses and brings their mind back to the here and now, offering a calming effect.

Project: Train a Support Team for Mental Health

Now let's take it one step further and put some of these ideas into action. Here's how to train a support team in your community to address mental health during a crisis. Having a team ready to step in and provide psychological first aid can make all the difference in a disaster situation.

Step 1: Select Team Members

Choose people who are empathetic, good listeners, and have a strong sense of calmness under pressure. You don't need mental health professionals (though they're invaluable), but selecting people who have experience with crisis situations or are compassionate is key.

Step 2: Provide Basic Training

Train your team in basic psychological first aid (you can find online courses or free resources from organizations like the Red Cross). Key areas that I would cover are:

- Active listening techniques
- Crisis de-escalation strategies
- Grounding exercises and mindfulness techniques
- Identifying signs of distress (e.g., withdrawal, anxiety, panic)
- Referrals for professional mental health support

Step 3: Set Up Communication Channels

Create clear communication channels for your team to use when needed. Have a list of emergency contacts, check-in points, and any resources available for referrals. Consider using walkie-talkies, community radios, or apps that work offline in case of power loss.

Chapter 3: Maintaining Positivity in Crisis

We've all heard the phrase "mind over matter," but when a real crisis hits. Staying positive can feel like an impossible task. Yet, the truth is that maintaining a positive mindset during tough times isn't just helpful. It's actually necessary for your survival.

How to Stay Positive and Focused in a Survival Situation

I'll be honest: staying positive during a crisis isn't always easy. There's nothing glamorous about running out of water during a power outage, trying to make a pot of soup with only canned beans and rice, or watching your neighborhood become unrecognizable after a storm. I've been there. But after going through a few situations where everything seemed to fall apart, I realized something important: attitude really is everything.

The first step is to remind yourself that this is temporary. Whether you're dealing with a severe disaster like an EMP attack or an economic downturn or a natural disaster, for instance, everything will eventually change. But when you're deep in the middle of a crisis, the only thing you can control is how you respond to it.

Focus on what you *can* do rather than what you can't. Can't access a store? Fine, figure out what you have in your pantry and make it work. Can't drive because there's no gas? Okay, how can you get from point A to point B without your car? It's about shifting from a mindset of helplessness to one of resourcefulness.

I remember a couple of years ago, my family and I were stuck without power for a few days after a severe snowstorm hit our area. We were huddled together with blankets, trying to stay warm, and the days stretched on longer than I'd like to admit.

At one point, we ran out of propane for the generator, which meant we were also without heat. I'll be honest, it felt like the walls were closing in. But then, we did what we always do: we turned it into a game. We started telling stories by candlelight, played card games to pass the time, and I put together a little scavenger hunt for the kids to find emergency supplies around the house. That change in perspective helped us all stay grounded and positive through an otherwise frustrating situation.

Using Routine and Self-Care to Boost Morale

In times of crisis, keeping some sense of normalcy is vital. This is where routine and self-care become the unsung heroes of survival. When everything feels unpredictable, having a basic routine can give you a sense of control, which helps reduce stress and boosts your morale.

Even if your routine looks a little different in a crisis, the act of sticking to something familiar can provide comfort and stability.

Build a Simple Daily Routine

When your world is turned upside down, it's easy to feel like everything is chaos. But maintaining a simple daily routine helps you regain a sense of normal amidst the madness. This doesn't need to be elaborate; just a few key things to anchor your day.

For example, we set a daily routine during that snowstorm. It wasn't about doing everything perfectly, but about creating small milestones that gave us a feeling of structure. We'd wake up, have breakfast, and then spend time checking on our neighbors or working together on household chores. The routine of it all (no matter how small) kept us grounded.

Practice Self-Care

When you're in crisis mode, self-care might seem like a luxury you can't afford. But here's the thing: you need it. If you don't take care of yourself (both physically and mentally) then your ability to help others or keep things going diminishes.

This doesn't have to be a whole spa day or anything extravagant. Simple actions like drinking enough water, eating something nourishing, taking a 10-minute walk, or even just finding a few minutes to breathe deeply and relax can make a big difference!

Project: Set Up a Family Routine for Crisis Management

Now that we've discussed the importance of maintaining positivity and self-care, it's time to put it into action. Your family's mental and emotional resilience during a crisis can be just as important as your physical survival skills. By setting up a simple family routine, you can help everyone stay focused, connected, and prepared. Here's a basic project outline to get you started:

Step 1: Create a Family Schedule

Sit down as a family and come up with a daily routine that everyone can follow during a crisis. This can include basic tasks like:

- Make sure everyone is accounted for, healthy, and has everything they need for the day ahead.
- Assign simple tasks (like helping with food prep, gathering firewood, or keeping watch) to keep everyone busy and involved.
- Set aside time each day for group activities like games, storytelling, or outdoor walks. This keeps morale up and ensures that the family stays emotionally connected.

Step 2: Build in Time for Self-Care

Make sure that everyone has personal time to unwind and take care of themselves, even if it's only for a few minutes. This can be something as simple as:

- Reading or journaling to process emotions
- Stretching or meditating to reduce tension
- Talking to a friend or loved one to stay socially connected

Step 3: Create Emergency Action Plans

In addition to your daily routine, discuss and prepare for potential scenarios that could happen during a crisis. For example:

- Where will you meet if you need to evacuate the house?
- How will you communicate if phone lines or the internet go down?
- Who is responsible for securing important documents or supplies?

Key Points To Remember:

1. Mental strength is just as important as physical preparation, as it will help you stay calm, focused, and in control during a crisis.
2. Building psychological resilience takes time and practice since it's about staying present, embracing a survival mindset, and being adaptable to the unexpected.

3. Having coping mechanisms like breathing exercises, breaking tasks into smaller steps, and staying connected with others can help you manage stress and stay grounded.
4. Helping others cope with stress requires active listening, offering reassurance, and using grounding techniques to bring them back to the present moment.
5. Staying positive and focused in a crisis is all about shifting your mindset from helplessness to resourcefulness, focusing on what you can do instead of what you can't.
6. Routine and self-care are vital for boosting morale and reducing stress, so create simple daily routines and make time for small acts of self-care.

BOOK 23:

PERSONAL DEFENSE AND WEAPONS

Chapter 1: Mastering Self-Defense

When it comes to surviving a crisis, it should go without saying that protecting yourself and your loved ones should be a top priority. But self-defense isn't just about fighting back. It's also about knowing how to stay calm, use your surroundings, and keep control of the situation.

Essential Hand-to-Hand Combat Techniques

Self-defense doesn't require you to be a martial arts expert, but it does require you to know the basics and to be able to use your body effectively to protect yourself. The primary hand-to-hand combat techniques that I would suggest learning are:

The Basic Stance

Before you even think about throwing punches or making moves, you need a solid foundation. The basic stance allows you to stay balanced and ready for anything.

- Stand with your feet about shoulder-width apart to maintain balance.
- If you're right-handed, your left foot should be slightly ahead of your right foot. This keeps you grounded and allows you to move quickly if you need to.
- Keep your hands up, not only for defense but also to protect your face. Make sure your dominant hand is by your chin, ready to throw a punch if necessary.

The Jab

The jab is a fundamental strike used to create distance and disrupt an attacker's momentum. Here's how to use it:

- Extend your lead arm (left if you're right-handed) straight forward while keeping your other hand up for defense.

- Aim for the attacker's face or chest; the goal is to create distance so you can follow up with a more powerful move or retreat.

The Elbow Strike

If you're in close quarters, your elbows can be your best weapon. The elbow strike is devastating and can stop an attacker in their tracks.

- From the guard position, rotate your hips and drive your elbow toward your attacker's face, jaw, or chest.
- The elbow strike is effective in tight spaces where a punch might not land.

Knee Strike

The knee strike is another close-quarters technique designed to cause maximum damage to an attacker's midsection, groin, or abdomen.

- Grab the attacker's head or shoulders (if you're close enough), and drive your knee upward into their midsection or groin.
- A good knee strike can incapacitate someone long enough for you to escape.

Defending Against Multiple Attackers

Handling a situation with multiple attackers is a whole different ball game. In reality, it's extremely challenging to fight multiple people at once, and survival should always be your top priority.

Your best chance is to create distance, use your surroundings, and incapacitate one person quickly to reduce the number of threats. You can do that with the following techniques:

Positioning Yourself

When facing multiple attackers, position yourself so that you can see all of them at once. Stand with your back against a wall or barrier if possible, so you have fewer directions to watch. Don't let yourself get surrounded, and try to position yourself in a way that prevents attackers from surrounding you. If they start to flank you, keep moving to stay in a defensible position.

Target the Weakest (If You Can)

When facing multiple attackers, you can't afford to waste time. Focus on one person at a time, and choose the weakest or closest person to incapacitate first. Strike quickly and aggressively, aiming for the vulnerable areas like the eyes, throat, elbow, knee, or groin.

Escape Over Confrontation

If you can, escape the situation. Your goal in a multi-attacker scenario should always be to create space and get away as quickly as possible. If you can reach a door, a public area, or even a car, run for it. Use any advantage (whether it's height, objects around you, or distractions) to give you the upper hand and make your getaway.

Use Your Environment

In an ideal world, you're not fighting in an open area. Use objects around you (such as chairs, tables, sticks, or even your bag) to defend yourself and create space. Throwing objects, if possible, can also disrupt your attackers' focus and give you a chance to act.

Chapter 2:
Firearms and Ammunition for Defense

When the grid goes down and raiders and marauders are taking to the streets, firearms can be one of the most reliable ways to protect yourself and your loved ones.

Basic Firearm Safety and Handling for Beginners

The first step in using firearms is always safety. It doesn't matter whether you're using a firearm to defend your home during an EMP event or are simply keeping yourself safe after a hurricane disrupts the power grid. That's because knowing how to handle your weapon properly is super important to avoiding accidents.

The 4 Cardinal Rules of Firearm Safety

1. Treat Every Firearm as If It's Loaded
2. Keep the Muzzle Pointed in a Safe Direction
3. Keep Your Finger Off the Trigger
4. Be Aware of Your Target and What's Beyond It

Storing and Managing Ammunition for Survival

Ammunition is just as important as the firearms themselves, especially when you're preparing for long-term survival. After an EMP attack or natural disaster, the last thing you want is to be caught without enough ammo to defend yourself or hunt for food. That's why properly storing and managing your ammunition ensures that it remains effective when you need it most.

Here are the most important tips to remember when storing ammo:

- Ammunition should be stored in a cool, dry, and dark place. Humidity can cause ammo to corrode or become ineffective, which can be disastrous in a long-term survival scenario.

Store your ammunition in sealed containers to protect it from moisture and temperature extremes.

- Invest in ammo cans or sealed plastic bags with desiccant packs to keep your ammo dry. I learned this the hard way after storing some in a garage and watching it corrode in high humidity during the summer months.
- If you're storing large amounts of ammunition, it's important to organize it by type, caliber, and purpose (self-defense, hunting, etc.). This will save you time when you need to quickly grab what you need in a post-disaster situation or when you're facing a multiple attacker situation.
- Ammunition does have a shelf life, and it's a good idea to rotate your stock. This ensures that older ammo gets used first, and your newest rounds stay fresh and effective. In times of crisis, this will help you manage your supply effectively, especially if you're uncertain when you'll be able to resupply.

Chapter 3:
Non-Lethal Defense Strategies

When it comes to protecting yourself in a survival situation, non-lethal tools can offer a smart and effective way to defend yourself without causing permanent harm. These are good weapons to have on hand as alternatives to firearms.

Using Pepper Spray, Tasers, and Other Non-Lethal Tools

Non-lethal tools like pepper spray and tasers can be incredibly useful for self-defense. In situations where you don't want to cause permanent injury but need to neutralize a threat quickly, these tools are a great option.

Pepper Spray

Pepper spray is a powerful and easy-to-use self-defense tool. It causes intense irritation to the eyes, throat, and skin to temporarily incapacitate an attacker.

Hold the can about 3 to 6 feet from the attacker, and aim for the face. Most pepper sprays have a range of up to 10-12 feet, which is enough distance to keep yourself safe while delivering a powerful blast.

Tasers

A taser delivers an electric shock that can temporarily incapacitate an attacker by disrupting their nervous system. Unlike pepper spray, tasers allow you to defend yourself from a greater distance and are effective against both individuals and animals.

Aim for the chest or abdomen area, as this provides the most effective contact. Once deployed, the prongs will deliver an electric shock, which will immobilize the attacker for several minutes.

Setting Up Non-Lethal Traps for Protection

While tools like pepper spray and tasers are great for direct defense, setting up non-lethal traps around your property can also provide protection.

Examples of non-lethal traps include:

- Simple Tripwire Traps
  - These can be used to alert you to intruders without causing harm. Set a tripwire connected to a bell or a noise-making device. When someone crosses the wire, it will trigger the alarm, giving you time to react.
- Net Traps
  - You can set up a net trap using sturdy rope or material designed to entangle an intruder, slowing them down long enough for you to assess the situation. While not a permanent solution, it can buy you valuable time.
- Pepper Spray Traps
  - For an added layer of security, you can rig pepper spray traps to go off when the wire is triggered. These traps spray an area with pepper spray, which will stop an intruder dead in their tracks without causing long-term harm.

Key Points To Remember:

1. Self-defense isn't just about fighting; it's about staying calm, using your surroundings, and keeping control of the situation in a crisis.
2. Mastering basic hand-to-hand combat techniques, like the jab, elbow strike, and knee strike, can make a huge difference in protecting yourself in close quarters.
3. In a multiple attacker scenario, your goal is to create distance, target the weakest attacker, and escape if possible – survival is always the priority.
4. Firearm safety is crucial, so always follow the 4 cardinal rules: treat every gun as if it's loaded, keep the muzzle pointed safely, and be aware of your target.
5. Properly storing ammunition in cool, dry places and rotating your stock ensures that it stays effective when you need it most.
6. Non-lethal tools like pepper spray, tasers, and simple traps (like tripwires or net traps) can be effective alternatives for defending yourself without causing permanent harm.

BOOK 24:
SURVIVAL FOOD FROM THE WILD

Chapter 1:
Identifying Edible Wild Plants

There's something magical about foraging for wild plants. It's like unlocking nature's pantry!

Foraging for Wild Plants and Mushrooms

If you've never foraged before, you might be surprised at just how many edible wild plants are out there. It's a skill that gets better the more you practice, and it's a great way to supplement your food supply.

Where to Look

Start by identifying common areas where wild plants grow. You don't need to be in the wilds of the forest; they can be found in urban areas, parks, fields, and even your own backyard. For example, dandelions (those bright yellow flowers people often treat as weeds) are actually edible! The leaves, roots, and flowers can all be eaten, and they're packed with nutrients.

Examples of common wild edible plants are:

- Dandelions
    - As I mentioned above, these are everywhere, and every part of the plant is edible. I often use the leaves in salads or cook them like spinach, the roots can be roasted as a coffee substitute, and the flowers are perfect for making wine.
- Plantain
    - Not the banana-like fruit, but a common green that grows in lawns and sidewalks. The leaves are edible and can be eaten raw in salads or cooked. Plantain is known for its medicinal properties as well, great for soothing bug bites and minor wounds.
- Chickweed

    - This small, creeping plant grows almost anywhere. It has a mild taste, and both the leaves and stems are edible. I often toss a handful of it into my salad mix when I go foraging. It's a great nutrient boost!

- Mushrooms
    - Mushrooms can be tricky, but once you learn how to identify them, they're a delicious and nutritious addition to your meals. Morels, chicken of the woods, and oysters are just a few of the edible ones. Just remember to never eat wild mushrooms unless you're 100% sure they're safe. This is one area where misidentification can be dangerous!

I remember the first time I found morel mushrooms. It was spring, and I was wandering through the woods near my house, looking for something new. I spotted these funny and spongy mushrooms and had no idea what they were at first. After doing some research and confirming they were safe, I cooked them up for dinner. Nothing beats the taste of food you've gathered yourself!

How to Avoid Poisonous Look-Alikes

Foraging is a great skill, but you've got to be cautious because there are poisonous plants and mushrooms that look very similar to edible ones.

Here are my top tips to avoid poisonous look-alikes:

- Leaves and Flowers
    - When identifying plants, look at the size, shape, and arrangement of the leaves. Also, pay attention to the flowers. Many edible plants have specific types of flowers (like white or yellow clusters). Poisonous plants, on the other hand, often have highly toxic parts that you can recognize by their smell or color.
- Smell
    - A great trick to identify certain plants is by smelling them. If the plant has a bitter or off-putting smell, it's likely poisonous. I've learned to use my sense of smell a lot when foraging, especially when I'm uncertain.
- Mushrooms
    - With mushrooms, you can't rely solely on color. Some poisonous mushrooms look very similar to edible varieties. Pay attention to spore prints (the pattern left by mushroom spores) and the gills. Death cap mushrooms, for example, are highly toxic and closely resemble some edible species.
- Wild Garlic vs. Lily of the Valley
    - Both plants have long, green leaves, but wild garlic smells like (you guessed it) garlic, while lily of the valley has a sweet but toxic scent. I remember being taught to gently crush the leaves and smell them when I first started foraging. That trick alone saved me from picking the wrong plant.
- The "Three Leaves" Rule
    - You've probably heard the phrase, "Leaves of three, let it be." It refers to plants like poison ivy, poison oak, and poison sumac, which all have clusters of three leaves. When you're out in the wild, always make sure you're not grabbing one of these.

Chapter 2:
Using Wild Plants for Medicine

When the world turns upside down knowing how to use wild plants for medicine is a skill that can keep you healthy and your family well. Nature has provided us with countless healing plants, and foraging for them can be just as useful as gathering food (if not more so).

Creating Herbal Remedies from Foraged Plants

Foraging for medicinal plants is a rewarding practice that goes beyond just providing food.

Here are some common medicinal plants that you can find:

- Echinacea
    - Commonly used to boost the immune system, Echinacea is easy to recognize by its purple flowers. I've used it several times in the past to stave off colds, especially when I didn't have access to over-the-counter cold medicine during an extended power outage.
- Plantain
    - This common weed often gets overlooked, but it's an excellent plant for soothing insect bites, rashes, or skin irritations. You can crush the leaves and apply them directly to the affected area, or make an infusion to drink.
- Lavender
    - Known for its calming properties, lavender is a great choice for making teas or adding to a sleep remedy. After a stressful situation like a hurricane and days without power, I've found lavender essential oils to be perfect for relaxation, especially when there's no time to wind down.
- Peppermint

- Not just great for tea, peppermint is also used for relieving headaches, digestive problems, and muscle pains. I've carried peppermint essential oil with me during long hikes, and a couple of drops rubbed on my temples worked wonders when a headache threatened to ruin the day.

Making Tinctures, Salves, and Teas for Health

When you've foraged the plants you need, the next step is preparing them into effective medicinal forms like tinctures, salves, and teas.

Tinctures

Tinctures are concentrated herbal extracts that are typically made by soaking the herb in alcohol. These are great for preserving the healing properties of plants for long periods of time.

Here's how to make a tincture:

- Harvest the plant material and chop it up to increase surface area.
- Place it in a glass jar, filling it halfway.
- Pour alcohol (like vodka or rum) over the plant material until it's fully submerged.
- Seal the jar and store it in a cool, dark place for 2-6 weeks. Shake it daily.
- Strain the plant material out and store the liquid in small bottles for use.

Salves

Salves are topical remedies that are perfect for soothing skin conditions like cuts, scrapes, burns, and bruises. They're made by infusing the medicinal herb in a carrier oil, which is then mixed with beeswax to create a balm-like consistency.

Here's how to make a salve:

- Choose your herbs (chamomile, plantain, or calendula work great for skin applications.
- Infuse your chosen herbs in a carrier oil (like olive oil or coconut oil) by heating the oil and herbs in a double boiler for 1-2 hours.
- Strain out the plant material and return the infused oil to the pot. Add beeswax (about 1 ounce of beeswax per cup of oil) and heat until it melts.

- Pour the mixture into small containers and let it cool and harden.

Teas

Making herbal teas is the simplest and most enjoyable way to benefit from the medicinal properties of wild plants. Herbal teas are ideal for soothing ailments, relaxing, or just enjoying the natural flavors of the plants you've gathered.

How to make tea:

- Boil water and add 1-2 teaspoons of dried or fresh herbs, depending on the plant.
- Let it steep for 5-10 minutes and strain out the plant material.
- You can sweeten it with honey or add lemon for flavor.

Chapter 3:
Long-Term Foraging Strategies

Foraging is one of those skills that can keep you connected to the land around you, whether you're living off the grid or just looking to supplement your food supply during tough times. But if you want to be truly self-sufficient, you need to think about foraging as a long-term strategy, and not just a one-time activity.

Sustainable Foraging Practices

When you're foraging for wild foods, there are a few key things to keep in mind:

- Only take what you need. Leave enough of the plant behind so it can continue to grow and reproduce. A good rule of thumb is to never harvest more than 10% of a plant population in any given area.
- Before you start foraging, make sure you can positively identify the plant. Some plants have toxic look-alikes, and you don't want to make a mistake when your food source is on the line.
- If you find a plant that's particularly useful or abundant in your area, consider learning how to plant and cultivate it. This is a long-term way to ensure you have a steady supply of food, and especially in a post-disaster world when wild populations may become harder to find.

I remember the first time I went foraging for blackberries in a wild patch near my home. At first, I picked more than I needed, thinking I could store a huge stash. But as I started learning more about foraging, I realized the importance of leaving some behind so the plants could spread and regenerate for next season. It was a game-changer in how I approached foraging, especially when I had to rely on it during a long blackout period.

Preserving Wild Foods for the Long Term

Once you've gathered wild foods, you'll need to preserve them for future use. In a long-term survival scenario, preserving food ensures that you're not reliant on immediate foraging and gives you something to fall back on when times get tough.

There are a few simple ways to store wild food:

Drying

Dehydrating is one of the best ways to preserve herbs, fruits, and even some mushrooms. When food is dried, it's lightweight, long-lasting, and easy to store.

Hang herbs in a dry, warm place with good air circulation, or use a dehydrator for faster drying. For mushrooms or berries, you can also use an oven on a very low setting or a solar dehydrator.

I learned this the hard way after a hurricane left us without power for days. We had a great wild food supply, but not enough of it was preserved for the long haul. After that, I started drying mushrooms and wild herbs as soon as I found them. A year later, that saved us during another blackout.

Canning

Canning is an effective way to preserve wild foods, especially for things like berries, tomatoes, or wild vegetables. Once sealed, jars of canned goods can last for a year or longer.

Make sure you have a pressure canner and jars. For most wild fruits, you can cook the food down into jams, jellies, or preserves, and then seal them in jars. The vegetables should be blanched before canning to preserve their nutrients.

During long-term disasters, having a stocked pantry of canned wild foods ensures you won't be left scrambling. I've relied on jars of homemade wild berry jam and foraged tomato sauce more times than I care to admit.

Freezing

Some wild foods, like certain berries and leafy greens, freeze well and retain their flavor and nutrients.

Wash and prep the plants, and then lay them flat on a baking sheet to freeze before transferring them to a sealed container or bag. This prevents them from freezing in a big block and makes them easy to pull out in portions.

Wild spinach and wild berries are perfect for freezing, and I've used my freezer to store them when I had too much to eat in one sitting.

Project: Set Up a Foraging and Preservation System

Now that you know the basics of sustainable foraging and preserving wild food, let's create a system that will make this a long-term habit for you.

I would start by researching which plants are available in your area during each season. Create a simple calendar where you can track the plants you need to look out for, and note down the best times to harvest them. This will keep you on schedule so you're always prepared when the right plants are in season.

You don't need a lot of fancy equipment, but trust me when I say that a few key tools will make the process easier:

- A dehydrator for drying fruits, mushrooms, and herbs.
- Jars and a pressure canner for canning.
- A deep freezer if you have the space and power.

Make sure you have everything set up in advance. For example, I started keeping plastic bins for storing dried foods and a cool, dark place for jars of canned goods. Preparation is key, especially after a storm that might knock out power for a while.

Next I would designate a wild foraging zone near your home or on your property. This could be a wild patch of herbs or even a section of forest known for mushrooms. As you gather and preserve your wild foods, don't forget to rotate them. Use the oldest preserved items first and replenish them with new harvests.

Key Points To Remember:

1. Foraging for wild plants is a valuable skill and can be done in urban areas, parks, fields, or your own backyard, with common edible plants like dandelions, plantain, and chickweed.

2. When foraging, always be cautious of poisonous look-alikes and use tricks like smelling the plants or checking the flowers to ensure what you're gathering is safe.

3. Wild mushrooms like morels, chicken of the woods, and oysters are delicious, but always double-check identification, as some toxic mushrooms look similar to edible ones.

4. Medicinal plants like Echinacea, plantain, lavender, and peppermint can be used to treat common ailments, making foraging just as useful for health as it is for food.

5. Preserve wild foods for long-term use by drying, canning, or freezing, ensuring you have food when foraging is no longer an option.

BOOK 25:

COMMUNITY DEFENSE STRATEGIES

Chapter 1: Organizing Community Defense

When a crisis strikes, your first instinct might be to protect your family. But the truth is, community defense is even more powerful. When you work together as a group, you can achieve much more than you could ever accomplish on your own.

Group Tactics for Securing Shelters and Resources

Once you've assembled a community defense team, the next step is to use group tactics for securing your shelter and your resources.

Securing Shelters

The first priority when setting up a community defense is to ensure your shelter is safe and easily defendable. You can use natural barriers like rivers, hills, or dense forests to your advantage if you're in a rural area. But in an urban environment, you'll have to make your shelter defensible by reinforcing windows, doors, and any entry points. I would give strong consideration to using fortified barricades or simply reinforcing doors with sturdy wood or metal if you're in a vulnerable area.

During an extended power outage after a storm, we turned our living room into a safe room by locking the windows, blocking any entryways with furniture, and setting up an area for sleeping and resting. We kept it simple (nothing fancy) but it gave us a central and safe place that we could defend if anything went wrong.

Group Patrolling and Watch Shifts

Setting up patrols with clear communication is key. If your team is well-coordinated, you can monitor the perimeter of your area to detect any threats before they reach your shelter. Establishing shift rotations is vital so that no one gets worn out or caught off guard. This allows you to keep watch around the clock.

If you're facing multiple attackers or looters, keeping an eye on your surroundings can give you the upper hand. We used walkie-talkies during that power outage to communicate with the other members of the team. Simple but effective, it allowed us to coordinate patrols, talk to neighbors, and quickly respond to any suspicious activity without wasting precious time.

Securing Resources

Resources like food, water, and medical supplies are the lifeblood of your defense team. You'll need to gather as much as you can early on and secure these resources so they don't become vulnerable targets. Think of your resources as your lifeline because without them, you risk running into major problems when supplies start running low.

During the same blizzard event, we established a centralized area to store food and water, with specific roles for community members. Everyone contributed what they had, and we made sure everything was rationed fairly. It helped ensure that no one hoarded supplies, and we all had access to what we needed when things started to get tense. Having these resources secured gave us the peace of mind to focus on other threats and challenges.

Project: Create a Community Defense Plan

Now that you've learned the basics of organizing your community for defense, it's time to put together a solid community defense plan. Start by drawing up a simple map of your community. It doesn't matter whether it's a few houses, an entire neighborhood, or a broader area.

Once your map is ready, it's time to assign roles based on each community member's skills. Things I would cover include:

- Watch and patrol shifts
- First aid and medical duties
- Food and water distribution
- Defensive positions or even negotiation in case things get tense with outsiders

Clear communication will be crucial in keeping everyone on the same page. Use walkie-talkies, two-way radios, or even signal flares if needed to stay in contact. I would also establish a group chat if possible (if you have communication systems available).

Your defense plan is only effective if it's practiced regularly. Schedule drills where your team practices securing the shelter, handling supplies, and communicating during an emergency. This will help your group stay organized and prepared for whatever comes next.

During a local flood a few years ago, we practiced our plan by setting up a mock emergency and running through it. We assigned people to various roles (patrol, supplies, medical, etc.) and practiced as though the flood was already underway. It was a great way to work out the kinks and make sure everyone was on the same page.

Chapter 2:
Building a Community Watch System

Having a network of eyes and ears can give you a major advantage in monitoring your surroundings and responding quickly to threats.

Setting Up Surveillance and Security

Surveillance systems don't have to be complicated, expensive, or rely on power. Start by looking for natural vantage points and strategic locations to set up surveillance. If you live in an urban area, rooftops or even higher floors can provide a great view of the streets below. If you're in a more rural setting, setting up lookouts along paths or around borders can give you an early warning system when people are approaching.

I remember setting up a simple surveillance system during a winter storm a few years ago. Our power had gone out, and we knew there were going to be some attempts at looting in the area. We used a combination of flashlights, mirrors, and even a few old walkie-talkies to keep an eye on entryways and make sure everyone stayed safe. It didn't require much, but it gave us a sense of control and security while the storm raged outside.

Using Drones for Patrol and Defense

Drones have become an increasingly popular tool for surveillance, and for good reason. They can give you a bird's-eye view of your entire area, scout out potential threats, and even respond to certain security situations in real-time.

I remember using a drone a few times during a local blackout after we heard rumors of people trying to break into homes. We sent up the drone to fly over the neighborhood and quickly spotted a group moving toward the edge of our community. By catching them early, we were able to set up a coordinated response, keeping everyone safe and avoiding any confrontation.

Here's how you can utilize drones for self-defense:

- Send drones ahead of any group patrols to check out the situation before anyone enters a potentially dangerous area. Drones can also help you track suspicious vehicles or individuals.
- Just the presence of a drone can act as a deterrent for anyone thinking about approaching your community with hostile intent. Knowing they're being watched is enough to make some people turn away.
- If you have multiple drones or multiple operators, drones can act as a communication tool by signaling and helping direct people to safe areas or alerting them of potential threats.

Project: Install a Basic Surveillance System

Now that you have a sense of how surveillance and drones can help, it's time to build your own basic surveillance system. You don't need to have a huge budget or the latest technology to set up something that will make a big difference for your community's security. Here's how to get started.

I would start out by identifying the most vulnerable spots in your community. This might include:

- Entrances to your neighborhood or property
- Key resource areas, like food storage or medical supplies
- High-traffic areas where movement might indicate a threat

Then draw a simple map of your area, and mark where you'll place cameras, lookouts, or other surveillance tools.

Depending on your budget, you can use a mix of cheap cameras, mirrors, and motion-sensor lights. Here's a basic list of what you'll need:

- Cameras (even basic security cameras will work)
- Flashlights or headlamps for signaling
- Walkie-talkies for communication between team members
- Reflective material or mirrors for spotting movement at a distance
- Drones (if available) for aerial surveillance

For our system, we started with basic cameras that we could attach to trees or utility poles. We also repurposed old cellphones as makeshift cameras that we could link to a network, and that worked surprisingly well when combined with simple motion-detection apps.

Install the equipment based on your earlier plan. Start with the most critical areas, such as the front gate, back doors, or entryways. If you're using cameras, make sure they're placed high enough to avoid tampering. For drones, you'll need to set up charging stations and make sure your team has enough experience with them to deploy quickly.

Chapter 3:
Defending Against Armed Attacks

No one likes to think about the possibility of armed intruders, but the reality is that it's something you need to prepare for. Whether you're living in a more rural area or stuck in a post-disaster urban environment, you could find yourself in a situation where you need to defend your home or your loved ones from someone who's willing to use force.

A few years ago, I had a wake-up call during a hurricane evacuation when things went south. The storm had knocked out power, and after a few days, a group of people started getting aggressive about scavenging for supplies. I had never been in a situation where I felt my home might be breached, but that week, we had to be extra vigilant. The thought of protecting my family (and not just protecting, but actually using physical force) became very real.

Using Defensive Tactics Without Lethal Force

There are times when lethal force isn't necessary or even the right choice. Here are some non-lethal tactics that you should have at your disposal:

- Using Barriers
    - If you can't get away from the attacker, barriers like furniture or any sturdy objects can be a lifesaver. When I had to deal with an aggressive person trying to break into my house after the flood, I used my kitchen table to block the door. It wasn't an elegant solution, but it gave me enough time to think and figure out what to do next.
- Verbal De-escalation
    - Sometimes, the best tactic is to talk. If you can see that the person is desperate but not yet fully committed to violence, you can try to de-escalate the situation by talking calmly, offering them food or water, or negotiating in a way that buys you time.
- Distraction and Escape

- If there's no way to fight without risking escalation, creating a distraction and then escaping might be your best move.

Key Points To Remember:

1. Community defense is powerful since working together as a group strengthens your ability to protect your shelter and resources more effectively than going solo.
2. Securing your shelter involves reinforcing entry points and using natural barriers like rivers or forests if possible, to create a defensible position.
3. Setting up group patrols and watch shifts with clear communication is essential to monitor your surroundings and stay ahead of potential threats.
4. Resources like food, water, and medical supplies must be gathered early and secured to ensure your community's survival during tough times.
5. Drones are a valuable tool for surveillance, providing real-time intelligence, acting as a deterrent, and assisting in coordinating responses during a crisis.

BOOK 26: SURVIVAL TRAVEL AND TRANSPORTATION

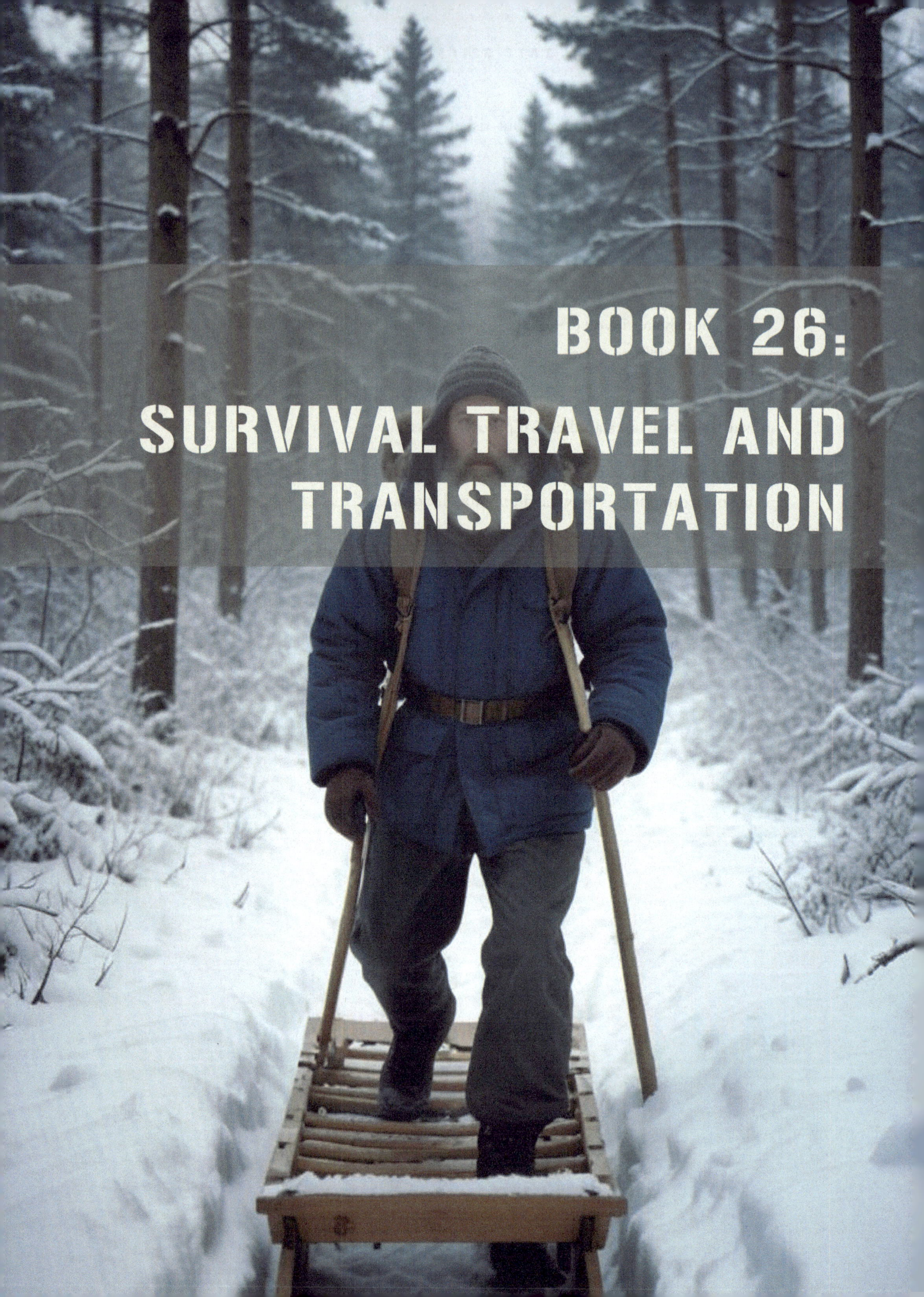

Chapter 1:
Off-Grid Travel Solutions

When the grid goes down, transportation automatically becomes a major issue. The good news is that there are ways to get around without relying on cars, trucks, or public transportation.

Bicycle-Powered Transport for Long Distances

If you've never considered a bicycle as a serious mode of transportation, it's time to start thinking about it. When gas stations are empty and vehicles are stuck in traffic, a bicycle is a reliable and efficient way to get around. Bicycles can cover long distances, don't require fuel, and can be maintained and repaired relatively easily. Plus, they're perfect for traveling off-road and getting into areas that would be hard to access by car.

I remember using my bike more than ever after a hurricane knocked out power in our neighborhood. With the streets blocked and everyone scrambling to find resources, my bike was a game-changer. I was able to pedal to the next town over, gather supplies, and return home without worrying about gasoline shortages or traffic jams. It wasn't the easiest ride, but it got me where I needed to go, and it's that kind of practicality that makes bicycles a great off-grid solution.

Using Horses and Livestock for Travel

When bicycles aren't enough, or you need to cover even more ground, horses and livestock can be the perfect solution. These animals have been used for centuries for travel and transport, and they're still one of the most effective ways to get from point A to point B when the roads are out or cars are no longer an option.

If you're thinking of using horses, there are a few key things you'll need:

- You'll need to train your horse (or other livestock) for travel. Horses are typically easy to train for riding and pulling carts, but it's a good idea to start small and work your way up.

- Horses can carry heavy loads, and you'll need the right equipment for carrying supplies. A pack saddle allows you to secure panniers or bags on the horse's sides to transport everything from food to gear.
- Don't forget that horses need food, water, and rest. Plan ahead for where you'll stop to feed them or where you'll find pasture for grazing.

Chapter 2:
Navigating Without Maps or GPS

When the grid goes down, you may also find yourself needing to navigate without the help of GPS or a map. But don't worry, because the skills to navigate using the sun, stars, and landmarks are timeless…and in my opinion they're also fun to learn!

Using the Sun, Stars, and Landmarks for Navigation

The world may seem uncertain in the aftermath of a disaster, but the natural world can offer us plenty of tools for finding our way.

The Sun for Direction

One of the simplest and most reliable ways to find your way during the day is by using the sun. The sun rises in the east and sets in the west, which provides a reliable east-west direction. If you know where the sun is during the day, you can get a general sense of which way you're heading.

One of the easiest ways to use the sun for navigation is by using a stick and its shadow. Place a stick in the ground, and mark the tip of its shadow. After about 10 minutes, mark where the shadow has moved. This marks a west-east line, with the shadow tip pointing west at midday and the stick pointing east.

The Stars for Nighttime Navigation

When the sun sets, it's easy to feel lost, but the stars can guide you, especially the North Star (Polaris) if you're in the Northern Hemisphere. Polaris is nearly directly above the North Pole, and it stays in a fixed position in the sky, making it a great reference point for finding north.

To locate Polaris, use the Big Dipper constellation. Find the two stars that make up the outer edge of the Big Dipper's "bowl" (these are called the pointer stars). Draw a straight line from these stars, and it will point directly to the North Star.

Landmarks for Orientation

Landmarks are another reliable way to navigate. If you're in a familiar area or you've passed through it before, you can use prominent features like mountains, rivers, or buildings to guide your way. If you're in unfamiliar terrain, you can mark your path using identifiable trees, rocks, or even man-made objects.

During a recent hiking trip, I had to rely on visible landmarks to retrace my steps after we got turned around. We identified a large ridge in the distance and used it as our point of reference, which made finding the trail much easier.

Techniques for Dead Reckoning and Pathfinding

Now that we've covered the basics of using natural elements for navigation, let's dive into a couple of more advanced techniques that are essential for survival: dead reckoning and pathfinding.

Dead Reckoning

Dead reckoning is a technique used to navigate when landmarks are unavailable. Essentially, it's a method of estimating your current position by using your last known position, speed, and direction. It sounds complex, but it's surprisingly effective when you're lost.

Start by determining your starting point. As you move, keep track of the distance you've traveled and the direction you've been heading. Use the sun, your watch, or a compass to help keep track of direction. If you're walking or hiking, count your steps to estimate how far you've traveled, or use pacing (like measuring how far you walk in 100 steps) to estimate distance.

Pathfinding

Pathfinding is about finding the most efficient way to move through difficult terrain, and it's a critical skill when there are no well-marked roads or trails. The goal is to avoid getting lost or trapped while moving toward your destination.

Look for natural features that might offer an easier route, like river valleys, ridgelines, or even animal trails. Follow the path of least resistance. If you're navigating through dense woods, for example, move along a creek or stream, which often leads to an exit or larger body of water.3

Chapter 3:
Traveling Long-Distances After Collapse

When the systems we depend on (like electricity, transportation, and commerce) break down, the ability to travel long distances without modern conveniences will naturally become a critical skill for you to learn.

Packing and Carrying Supplies on the Move

If you're traveling long distances after a collapse, what you carry with you is just as important as how you carry it. When resources are scarce, you'll need to make sure that everything you bring is necessary, durable, and easily transportable. There's no room for unnecessary weight or items you won't use!

Start by focusing on the essentials like food, water, shelter, and tools. Here's a breakdown of what you'll need to have with you for long-distance travel:

- Food and Water
    - You'll need to bring enough non-perishable food to last you for the duration of your journey. Think about lightweight, high-calorie items like jerky, nuts, dried fruits, or energy bars. Water is even more important. I would consider bringing a portable water filter or purification tablets, as you'll likely need to source water from rivers or streams along the way.
- Shelter
    - A compact, lightweight shelter is crucial. While it's tempting to bring something heavy or fancy, your shelter should be small and easy to set up. A tarpaulin or lightweight tent is a good option for packing. In addition, include a sleeping bag or emergency blanket to keep warm at night.
- Clothing and Footwear

    - For long-distance travel, pack durable clothing suited to the weather conditions, along with sturdy boots. Layering is key to adjust to changing temperatures. When I was on a long hike after a storm cut off our access to the local roads, I wore multiple layers and good boots, which made all the difference in comfort and safety.

- Tools and First Aid
    - Always bring a multi-tool, a knife, a fire starter, and a small first-aid kit. These tools will serve a variety of needs, from food prep to making repairs, building shelters, and tending to injuries.

Pack your heaviest items close to your back, and lighter items at the top or in the front pockets. This helps maintain your balance while walking long distances. Keep frequently needed items, like snacks, a water bottle, or a map, in easy-to-access areas.

Finding Shelter and Resources While Traveling

Traveling long distances after collapse means that you'll need to be resourceful in finding shelter and resources along the way. You'll have to be flexible and adapt to the environment you're traveling through.

If you're traveling through urban areas, abandoned buildings can provide temporary shelter. In rural areas, you might need to rely on natural shelters like caves or dense thickets. Look for places that provide protection from the elements and from potential threats.

When you can't find an established shelter, your next best option is to set up your own. Look for natural windbreaks or create a makeshift shelter from a tarpaulin and branches. Try to stay hidden as much as possible to avoid attracting attention.

Along the way, you'll need to find resources like food, water, and firewood. Always be aware of water sources. Rivers, streams, and lakes are all potential resources, but make sure you filter or boil the water to avoid illness. While foraging can help supplement your supplies, you'll likely need to rely on whatever you've packed for the bulk of your meals. If you're traveling through an area with wildlife, you might be able to hunt small animals or fish for food, but make sure you know how to do this safely and humanely.

Project: Build a Portable Shelter for Travel

Now that we've covered the basics of traveling long distances and managing your supplies, it's time to build your own portable shelter for travel. A reliable shelter is one of the most important things you can have when you're out on the move, and it doesn't have to be complicated or expensive to build.

Step 1: Choose Your Shelter Type

There are several options for building a portable shelter. Here are a few ideas:

- Tarp Shelter
    - A lightweight tarp can be strung between two trees or poles to create a simple, waterproof shelter. You can use paracord to tie it down, and it's easy to set up and break down.
- Tent
    - A compact, two-person tent can offer more protection from the elements. It's heavier than a tarp, but it can provide better shelter if you're in harsh conditions.
- Bivvy Sack
    - A bivvy sack is an ultra-lightweight option, especially useful for one person. It's essentially a waterproof sleeping bag that gives you the basics of shelter when you don't have much space or weight to spare.

Step 2: Gather Materials

For a tarp shelter, you'll need:

- A tarp (at least 8' x 10')
- Paracord or strong rope
- Stakes (you can make these from branches if necessary)
- A sleeping bag or blanket

For a tent, make sure to pack your:

- Tent and stakes
- Sleeping pad or ground tarp for extra insulation
- Rain fly (to ensure waterproofing)

Step 3: Set Up Your Shelter

Once you've chosen your shelter type, practice setting it up before you need it. Choose a location with some shelter from the wind and elements, and try to elevate your shelter off the ground if possible. If you're using a tarp shelter, practice using triangular configurations or A-frame setups for the best protection.

Source

Key Points To Remember:

1. Off-grid transportation solutions like bicycles and horses can be incredibly effective when vehicles are unavailable, and they're easy to maintain and use for long-distance travel.
2. Using natural navigation techniques like the sun, stars, and landmarks allows you to travel without maps or GPS, providing a reliable way to find your direction.
3. For advanced navigation, dead reckoning (estimating your position based on distance and direction) and pathfinding (using natural features for easier routes) are essential skills to master.
4. When traveling long distances after a collapse, focus on packing only the essentials: food, water, shelter, clothing, tools, and first aid to ensure you can carry everything safely and efficiently.
5. Shelter and resources will be scarce, so look for abandoned buildings in urban areas or natural shelters like caves in rural areas, and be prepared to set up your own shelter when necessary.

6. A portable shelter is a must for long-distance travel, and options like tarp shelters, compact tents, or bivvy sacks can offer protection from the elements while remaining lightweight and easy to pack.

BOOK 27: FARMING AND LIVESTOCK CARE

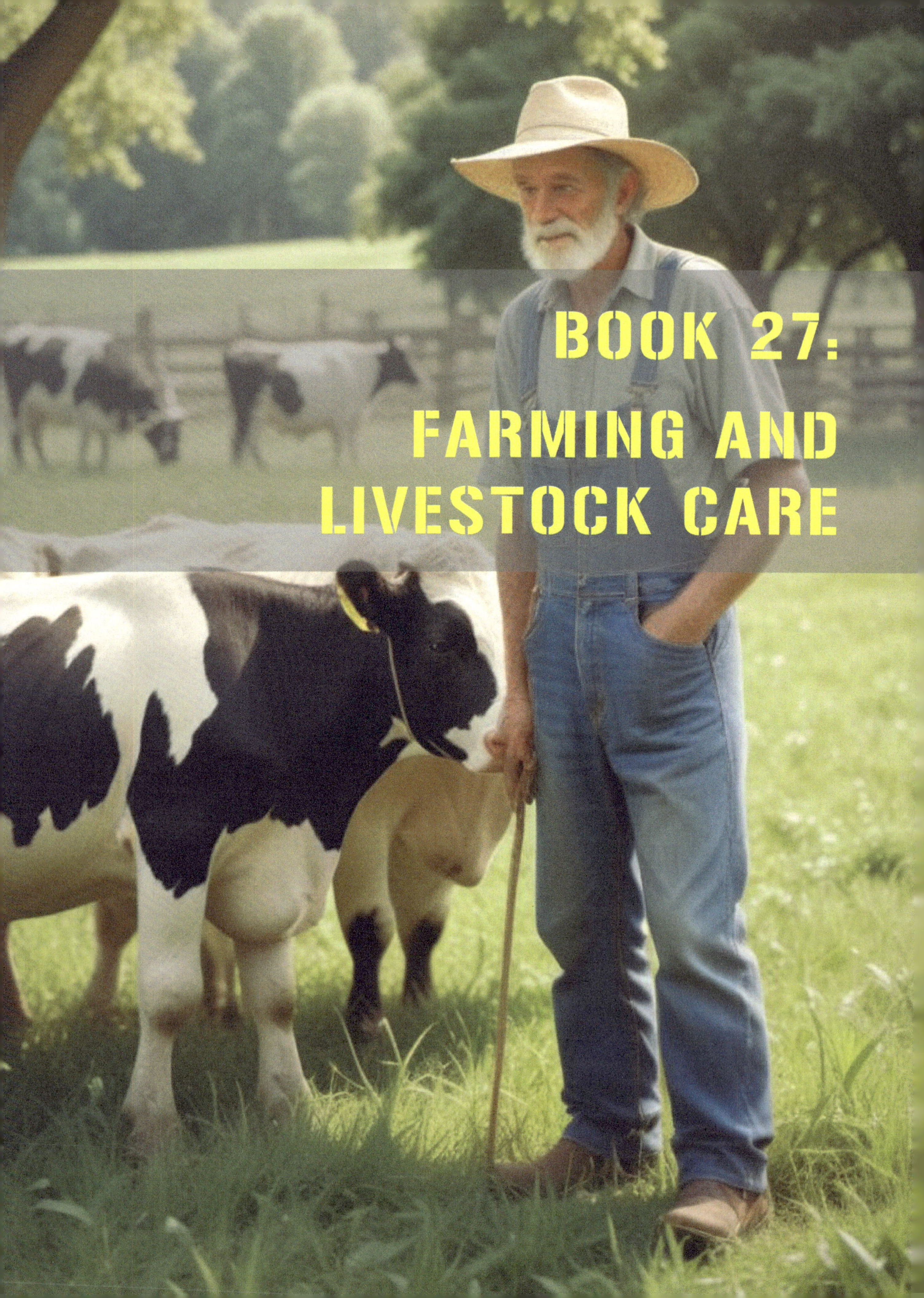

Chapter 1:
Raising Livestock for Self-Sufficiency

When the world around you starts to crumble, being able to provide for yourself becomes more important than ever. One of the best ways to ensure you're ready for any situation is to become self-sufficient in food production. Raising livestock can provide you with meat, milk, eggs, and even manure for your garden.

Caring for Chickens, Goats, and Pigs for Meat and Milk

Livestock are an excellent source of food, and when you're raising them for self-sufficiency, you need to think about how to get the most out of them. Chickens, goats, and pigs each provide different benefits, and each requires specific care to thrive in a home farm setting.

Chickens: Eggs, Meat, and Pest Control All At Once

Chickens are probably one of the easiest and most rewarding animals to raise. Not only do they provide eggs, which are a great source of protein, but they can also be butchered for meat. They also serve as natural pest control, keeping insects and weeds down around your property.

- Chickens need a safe coop to protect them from predators, as well as a secure run to roam freely during the day. Make sure they have access to clean water and food (grains, seeds, and scraps work well), and they'll need nesting boxes for laying eggs.

- Most chickens start laying eggs at around 5 to 6 months of age. You'll want to collect eggs daily to keep the nest clean and encourage more egg-laying. I've raised a small flock for a few years, and the steady supply of eggs has been a lifesaver during supply shortages after a storm or power outage.

- When it comes time to harvest meat, broilers (chickens bred for meat) are ideal, but any chicken can be used for meat once they're past their egg-laying prime.

Goats: Milk, Meat, and Fiber

Goats are incredibly versatile and can provide milk, meat, and even fiber, depending on the breed. Dairy goats, like Nubians or Alpines, can supply you with fresh milk daily, which can be used for drinking or turned into cheese, butter, and yogurt.

- Goats are hardy animals, but they do need a well-fenced area to roam and forage. They're natural browsers, meaning they prefer to eat shrubs, leaves, and weeds rather than grass. They need access to fresh water, a balanced diet (grain, hay, and browse), and protection from predators.
- A dairy goat can produce 1 to 2 gallons of milk per day during the peak of lactation. Regular milking is necessary to keep the milk flowing, and you'll need to pasteurize the milk if you're planning to drink it raw. Goat milk is rich in nutrients and can be turned into delicious cheese, which can be stored or used in recipes.
- If you raise goats for meat, the most common breeds are Boer goats or Kiko goats, both known for their fast growth and tender meat. Goats can be harvested for meat when they're 6 to 12 months old.

Pigs: Meat and Waste Reduction

Pigs are a great source of meat, and they can also help with waste management by eating food scraps and other organic material. They're hearty animals that grow quickly and provide a lot of meat in a short amount of time.

- Pigs need a clean, dry place to sleep, and plenty of space to root around for food. They're omnivores, so you can feed them a mix of grains, scraps, and foraged food. They also need fresh water daily.
- Pigs can be raised for bacon, ham, sausage, and other cuts of meat. Depending on the breed, pigs can be ready for slaughter in about 6-12 months. Large White and Landrace pigs are good for meat production.
- Pigs can be incredibly efficient in turning food scraps into meat. During a post-disaster scenario, I found that feeding pigs kitchen scraps and garden waste was an efficient way to reduce waste while also producing valuable protein.

Animal Shelters

Your animals will need a safe, comfortable, and secure environment to thrive, which means providing them with proper shelters and managing their health carefully.

Each type of animal will need its own specific type of shelter. Here are the basics:

- Chicken Coop
    - Chickens need a well-ventilated coop that's predator-proof. The coop should be dry, and it should have roosting bars for the chickens to sleep on. Make sure the coop is large enough for the flock to move around comfortably and have separate nesting boxes for the hens to lay eggs.
- Goat Shed
    - Goats need a secure shed that protects them from the elements. This shelter doesn't need to be fancy but should provide enough space for the goats to lay down and stand comfortably. Make sure it's well-ventilated and dry, with plenty of straw bedding to keep them warm in winter.
- Pig Pen
    - Pigs need a strong fence to prevent escape, as they're excellent at rooting under barriers. The pen should have a muddy area for them to wallow in, as this helps them cool down and prevent skin problems. They also need a dry area for sleeping, along with plenty of fresh food and water.

Chapter 2:
Sustainable Farming Techniques

When the world's going sideways you'll want to be able to provide for yourself without having to rely on grocery stores.

Crop Rotation and Soil Fertility Management

One of the first things I learned about gardening (the hard way, of course) is that soil isn't an endless resource. It gets tired, worn out, and, eventually, crops stop growing well if you don't treat it with care. Think of soil as a marathon runner. It can only go so far before it needs a break and a little TLC.

Crop Rotation: Giving Your Soil a Break

When you plant the same crops in the same spot year after year, the soil's nutrients get depleted. It's like eating the same meal every day. You eventually get sick of it, and your body starts asking for something else. Crop rotation helps prevent this. You switch up the crops you plant in each section of the garden to avoid depleting the soil of the same nutrients.

Here's how it works: let's say you plant tomatoes in one bed this year. Next year, instead of planting tomatoes there again, you could plant beans, which fix nitrogen in the soil, or root vegetables like carrots that don't need as much nitrogen. The idea is to give each type of crop its own year to rest while allowing the soil to regenerate. After a few years, you'll have a nice, nutrient-rich patch ready for your next round of tomatoes (or whatever crop you're craving).

I'll admit, the first time I tried crop rotation, I was too excited about the idea of planting all my favorite crops in every bed, every year. Guess what? It didn't work. My soil got depleted, and my crops grew half-heartedly at best. I remember planting tomatoes in the same spot for two seasons in a row. After that, they looked like they were auditioning for a zombie movie. I learned the hard way that crop rotation isn't just a nice idea—it's a must if you want to avoid constantly having to buy fertilizer or watch your veggies shrivel up in the dirt.

Soil Fertility: Giving the Earth What It Needs

Good soil is the foundation of any successful garden. And while crop rotation helps, sometimes you need to give the soil a little extra love. Here are a few ways to do that:

- Compost
  - Adding compost is like giving your soil a multivitamin. It replenishes the nutrients, adds organic matter, and helps your plants grow stronger. If you don't have a compost pile yet, get on it! You can use kitchen scraps (except for meat and dairy), yard waste, and even coffee grounds. The more, the better.
- Cover Crops
  - Planting cover crops like clover or alfalfa can help keep the soil covered, prevent erosion, and add nutrients back into the ground. It's like giving the soil a spa day while you take a break from farming.
- Mulch
  - A good layer of mulch keeps the soil moist, regulates temperature, and adds organic matter as it decomposes. Plus, it looks nice, like your garden's wearing a cozy blanket.

Companion Planting and Pest Control

Now, let's talk about making your garden a little less hostile and a lot more friendly. Companion planting is like creating a "buddy system" for your plants. Some plants grow better together because they help each other out. Others, well, they're like that one friend who shows up late, eats your snacks, and leaves a mess. We're aiming for teamwork, not chaos.

One of the best combos I've had in my garden is tomatoes and basil. These two are like peanut butter and jelly. They just work. Basil helps repel insects like aphids and mosquitoes, which can wreak havoc on tomatoes. Plus, the smell of basil seems to boost the flavor of the tomatoes. It's like they're making each other better. Talk about a power couple.

Other great pairings include:

- Carrots and Onions
  - The onions' scent helps keep pests away from the carrots. It's like they're giving each other a shield.

- Marigolds and almost anything
    - Marigolds are natural pest repellents, especially for nematodes, those nasty little critters that like to mess with your roots. Plant a few around your garden to keep pests in check.

Chapter 3: Integrating Livestock and Farming

When it comes to self-sufficiency, the magic happens when you can integrate livestock and farming into a harmonious and sustainable system.

Raising Livestock Alongside Crops for Sustainable Living

If you've ever raised animals and grown food, you know that one doesn't have to be separate from the other. In fact, integrating livestock and crops is one of the most sustainable ways to farm. The key is balance.

Here's how the cycle of life works in an integrated system:

- Manure as Fertilizer
  - Livestock, particularly cows, chickens, and pigs, provide rich manure that can be used to fertilize crops. When properly managed and composted, manure can become black gold for your garden, increasing soil fertility and helping crops grow strong.
- Pest Control
  - Chickens are natural pest controllers. They'll happily eat up insects, weeds, and even small rodents, which can be a nuisance in a garden. I've seen my chickens do wonders for my vegetable beds. I let them roam freely during the winter, and when spring rolled around, the ground was pretty well cleared of pesky weeds and bugs.
- Weed Management
  - Goats can help keep weeds in check around your fields and garden. They'll happily munch on anything green, keeping unwanted vegetation from overtaking your crops. However, make sure to give them a specific grazing area—you don't want them eating your plants! I learned this the hard way when I accidentally let my goats loose near the garden. They had a feast, and I had to replant half my crops.

- Draft Power
    - If you're lucky enough to have larger animals, like oxen or draft horses, they can be used for plowing, hauling, and other farm tasks, saving you a ton of energy and effort compared to relying solely on human muscle. When we first started integrating animals on our farm, my old horse, Buddy, was the key to moving heavy loads around, saving us time when we had to transport mulch, compost, and firewood.

Managing Manure and Fertilizer for Gardens

When you're raising livestock, manure is a given, and it can be a good thing for your garden. However, manure must be managed properly to avoid problems like burning your crops with fresh manure or attracting unwanted pests.

How to Compost Manure

Composting manure is the key to turning your livestock waste into nutrient-rich fertilizer for your garden. The basic principle is to mix the manure with carbon-rich materials, like straw, leaves, or sawdust, to ensure it breaks down properly without becoming too stinky or creating harmful bacteria.

Start with a layer of straw or wood chips at the bottom of your compost pile. Then, add layers of manure (chicken, goat, pig, etc.), followed by more straw or leaves. Turn the pile every few weeks to aerate it, and keep it moist (but not soggy). Over time, the pile will break down into dark, crumbly compost that's ready to use in your garden. It usually takes about 6 months to a year for manure to fully compost.

I learned the importance of properly composting manure the hard way one spring when I didn't let my chicken manure compost long enough before adding it to my soil. I thought, "Hey, more manure, more plants!" But I ended up burning a good portion of my garden with fresh chicken poop. Let's just say my tomatoes didn't exactly thank me for the "extra nutrients." I had to replant and learned that patience is key when working with manure.

Project: Build a Chicken Coop and Garden Combo

Now that we've covered how to integrate livestock and crops, it's time to get hands-on. The goal of this project is to build a chicken coop that works alongside your garden, providing your chickens with a safe place to roost while allowing them to help fertilize and clear your garden.

Step 1: Choose the Right Spot

Pick a spot that's near the garden but not too close. You want your chickens to be able to roam and help keep pests down but not destroy your garden beds with too much scratching and pecking. Make sure the area is well-drained and has access to sunlight for warmth and light.

Step 2: Build the Coop

You don't need a fancy coop. Just something that will keep your chickens safe from predators like foxes, raccoons, and hawks. Here's a basic outline:

- Frame
  - Use sturdy wood or recycled materials to create the frame of the coop. A simple rectangular design works well.
- Floor
  - Install a floor to keep the chickens safe from burrowing predators. A wooden floor is fine, but make sure to add shredded straw or pine shavings for bedding.
- Roosting Bars
  - Chickens need a place to roost, so install horizontal bars at different heights for the chickens to sleep on.
- Nesting Boxes
  - Add a few boxes (one for every 3-4 hens) for the chickens to lay their eggs in. I use simple wooden boxes lined with straw.

Step 3: Build the Run

Once your coop is built, create a run (a fenced-in area) where the chickens can roam during the day. A chicken run should have:

- Wire mesh (chicken wire) around the sides and top to prevent predators from getting in.
- Easy access for you to enter and collect eggs, clean out the coop, and check on the chickens.

Step 4: Integration with the Garden

Let your chickens roam in and out of the garden, helping to clear out weeds and eat pests. If you want to control their access, use portable fencing to move them around different parts of the garden. Over time, they'll provide natural fertilizer while doing a great job of keeping your garden healthy.

Source

Key Points To Remember:

1. Chickens are one of the easiest animals to raise, providing eggs, meat, and pest control for your garden, making them a great addition to any homestead.
2. Goats offer milk, meat, and fiber, and they're hardy animals that require good fencing and proper grazing management to thrive.
3. Pigs are efficient at converting food scraps into meat and are excellent for waste reduction, but they need space, a secure pen, and regular access to food and water.
4. Manure from livestock like chickens, goats, and pigs can be used as valuable fertilizer for crops, but it must be composted properly to avoid burning your plants.
5. Integrating livestock and crops creates a sustainable system, with animals like chickens and goats providing pest control, weed management, and natural fertilizer for your garden.

BOOK 28:
PREPARING FOR A GLOBAL CRISIS

Chapter 1:
Understanding Global Economic Crises

When the economic systems we depend on crumble, the effects can be felt on a massive scale. A global economic crisis can and will disrupt daily life in ways we can't predict. But here's the thing: even when the world feels like it's spiraling out of control, there are steps you can take to prepare and make sure you and your family are financially resilient.

How Economic Systems Collapse and Impact Society

Economic systems are like the backbone of society. They're what allow us to trade goods and services, pay for necessities, and ensure the smooth functioning of government. But when those systems start to break down, things can go south quickly.

A major economic collapse can happen for a variety of reasons, but let's look at the big ones:

- Currency Devaluation
    - Sometimes, governments print too much money, leading to inflation. When this gets out of control, it can result in hyperinflation, where prices skyrocket and money becomes nearly worthless. This is often the result of a failing economy, war, or massive government debt.
- Bank Failures
    - If banks collapse, people lose access to their savings. We saw this during the 2008 financial crisis, and in extreme cases, like during the Great Depression, people literally couldn't access their money. This can lead to a liquidity crisis, where cash is in short supply, and businesses struggle to keep running.
- Resource Scarcity

- A collapse in the global supply chain can lead to scarcity in essential resources like food, fuel, and medicine. When these basics become scarce, the economy struggles to function, and people begin hoarding what they can get their hands on.

Preparing for Hyperinflation and Resource Scarcity

So, now that we know the reasons why economic systems can collapse, how can we prepare for these worst-case scenarios? The key is to think about hyperinflation and resource scarcity in terms of survival. I'm talking about food, water, security, and maybe even currency replacement systems like bartering.

Hyperinflation: The Wallet-Killer

Hyperinflation is the point at which inflation goes wild. Prices rise uncontrollably, and your money's buying power shrinks drastically. Imagine buying a loaf of bread today for $2 and then paying $50 for that same loaf in just a few weeks. That's hyperinflation.

To prepare for this, here's what you can do:

- When currency is no longer reliable, assets like gold, silver, and other commodities become valuable. You don't have to go out and buy an entire vault of gold, but consider holding onto small amounts of precious metals or other barterable goods (think ammo, tools, or medicine).
- The last thing you want is to be stuck paying ridiculous prices for food. Stockpile food and water that will last months or even years. Focus on high-calorie, easy-to-store items like rice, beans, canned goods, and freeze-dried food. I learned this after a hurricane knocked out power for weeks—having that extra stash of rice and canned goods was a game-changer when the grocery stores were empty.
- Having all your money tied up in a single bank account or investment is risky in a collapsing economy. Diversify by considering property, hard goods, or even small-scale farming to ensure you have resources even if money isn't as useful.

Chapter 2:
Surviving Political and Social Unrest

When political systems break down or social unrest takes over, chaos tends to follow. Civil unrest, protests, and outright lawlessness can become a major threat to your safety and well-being.

Protecting Yourself During Political or Civil Crises

When chaos erupts in the streets, the primary goal is to stay safe and avoid unnecessary risk. Here's how you can prepare to protect yourself during a crisis:

The first step in staying safe during a political or civil crisis is knowing what's happening around you. Information is power, and the more you know, the better prepared you'll be. In times of unrest, news may not be readily available, or it may be unreliable.

It's wise to stay connected to multiple information sources, such as local radio stations, emergency alerts, or independent news outlets. In case the internet is down, consider keeping a battery-powered radio on hand to tune into emergency broadcasts.

When protests or violence begin, staying low and avoiding involvement is often the best way to stay safe. The streets can be full of anger and confusion, and even the most peaceful gatherings can turn chaotic. Stay away from any mass gatherings if you can. It's easy to get caught in the middle of something when the situation escalates quickly.

If you hear about protests or unrest in your area, stay inside. Close doors and windows, stay away from the front lines, and monitor the situation from the safety of your home. My neighborhood during a citywide protest in 2013 was quiet for a while, but as tensions rose, things began escalating outside. Staying inside kept us out of harm's way, and we were able to take care of essential needs from a safe distance.

Understanding the Role of Lawlessness and Disorder

When the law starts breaking down, lawlessness takes over. This is when society's norms and rules cease to function effectively.

The Collapse of Order

In a civil crisis, the legal system may become overwhelmed or completely broken down, making it harder to trust law enforcement to protect you. When this happens, your best defense is to protect yourself and rely on the resources and community support you've already set up.

In extreme cases, people may form vigilante groups or neighborhood watch systems to protect their homes and families. If the police are unavailable or compromised, these groups can serve as a last line of defense.

In times of lawlessness, community becomes everything. If you're part of a neighborhood watch system, or if you have close ties to neighbors, you can lean on each other for support, protection, and resources.

Project: Create a Neighborhood Watch System

It's one thing to protect yourself, but the best defense is often when your whole neighborhood is on the same page. Here's how to start a neighborhood watch system that keeps everyone informed, safe, and ready to react in times of crisis.

Find like-minded individuals in your neighborhood who are willing to help with security. You don't need to have a large group. Just a few committed people can make a big difference. Organize a meeting and discuss how you'll work together.

Assign different responsibilities based on skill sets. For example, if someone is trained in first aid, they can be responsible for health emergencies. If another person has military or security training, they can oversee the defense strategy. Everyone should know what role they're playing in the event of unrest.

Make sure that everyone in your neighborhood watch is easily reachable. Set up group chats, use walkie-talkies for emergencies, and make sure everyone knows the best way to communicate when cell networks fail. Keep a list of emergency numbers and contacts, and establish clear signals for calling for help or signaling danger.

Develop a neighborhood defense strategy, including safe zones and emergency exit routes. Discuss where people will go if things get out of hand and how they'll keep the area secure. Practice drills and make sure everyone knows the plan.

Key Points To Remember:

1. Economic systems can collapse due to currency devaluation, bank failures, or resource scarcity, leading to hyperinflation.
2. To prepare for hyperinflation, diversify your assets by holding precious metals, stockpiling food and water, and considering alternative forms of wealth like property or small-scale farming.
3. During political and social unrest, staying informed through multiple sources and avoiding mass gatherings are essential for personal safety.
4. When lawlessness prevails, community support and neighborhood watch systems become vital for protection and resource management.
5. Establishing a neighborhood watch system involves organizing a small group of committed individuals to oversee security, first aid, and communication during times of crisis.

BOOK 29: DIY DISASTER RECOVERY SOLUTIONS

Chapter 1:
Basic Home Repairs After a Disaster

When disaster strikes, the impact on your home and shelter can be catastrophic. But fear not! You don't have to be a professional contractor to get your home back in shape.

Rebuilding Damaged Homes and Structures

After any disaster, your first priority is to make sure your shelter is safe and livable again. Depending on the severity of the damage, the repairs you'll need to make could be as simple as patching up a few leaks or as complex as completely rebuilding part of your home.

Before you start swinging a hammer, take some time to properly assess the damage. Look for:

- Check the foundation, walls, and roof. If your home has experienced major structural damage, it may require professional help, but for less severe issues, you can usually tackle it yourself.
- If there's been a flood or heavy rainfall, check for water damage inside the home. Pay special attention to walls, flooring, and ceiling insulation.
- If you're dealing with an EMP attack or electrical failure, make sure the electrical system is safe to handle. If you're not confident about the wiring, it's best to call in an expert.
- Roofs are often the first part of the house to take a hit during a storm. If you've got a leaking roof, you'll want to address it quickly to avoid more water damage inside your home.

Temporary Fixes First

When the damage is significant, don't go straight for the complete rebuild. Start with temporary fixes that will allow you to stay safe until you can complete the long-term repairs.

- Tarping

    - For roof leaks, use a heavy-duty tarp to cover the damaged area. Secure it tightly with duct tape, nails, or wooden battens to keep water out until you can repair it properly.

- Boarding up windows
    - If windows were broken, use plywood or other sturdy materials to cover the openings. You don't want to leave your home exposed to the elements or burglars.

Once you've dealt with temporary fixes, it's time to focus on rebuilding key areas:

- If your foundation was compromised, you'll need to repair cracks or even reinforce it. If the damage is minor, you can often patch cracks with concrete patching compound. For more serious issues, you might need to consult a professional.
- If walls were damaged due to flooding or structural issues, you'll likely need to replace drywall or plaster. For simple repairs, you can replace sections of drywall and repaint once everything's dry.
- If the roof is damaged beyond temporary fixes, you may need to replace shingles, tiles, or even the entire roofing structure. Make sure to use waterproof materials when patching or replacing roofing sections.

Finding Alternative Building Materials

After a major disaster, regular building materials may be in short supply, and you'll need to get creative. Fortunately, there are many alternative building materials that can serve you in a pinch, and they often come from resources already around you.

Scavenging Local Resources

When traditional materials like lumber or bricks aren't available, look to your environment. Depending on your location, you may be able to:

- Salvage wood from old furniture, pallets, or even nearby destroyed homes.
- Use stones or bricks from collapsed walls or damaged structures.
- Repurpose metal scrap for roofing or framing. Be cautious when working with metals, and make sure to handle them safely.

- If you're building a more permanent structure and you're in a location with abundant dirt, you can create your own adobe bricks or use earthbags to build walls. These are materials that can be made on-site and are surprisingly effective for insulation.
- In certain regions, bamboo or reeds can be used for wall construction or even roofing. They're lightweight, durable, and easy to gather in the right climate.

Chapter 2:
Creating Off-Grid Solutions for Urban Recovery

When society faces collapse the urban infrastructure we rely on can quickly break down. Power goes out, water systems fail, and food becomes scarce. But here's the thing: you don't need to wait around for the world to reset. In fact, by implementing off-grid solutions, you can help make sure that you and your community have everything you need to survive and rebuild.

Implementing Off-Grid Power, Water, and Food Systems

Urban recovery in a post-crisis world doesn't have to mean total reliance on outside help. By shifting to off-grid systems, you can keep yourself and your community supplied with power, water, and food, even when the grid goes down.

Off-Grid Power Systems: Solar, Wind, and More

When the lights go out, it's easy to feel powerless (pun intended). But don't fret—there are several off-grid energy sources that can keep your home or community running without relying on the main power grid.

- Solar Power
  - Solar panels are one of the most accessible off-grid power solutions. If you have access to a roof or even an open courtyard, you can install solar panels to generate electricity. With a decent-sized solar array, you could power everything from lights to small appliances to refrigeration (which, as you'll see, is key to food preservation).
- Solar Batteries
  - To make sure you have power when the sun isn't shining, you'll need a battery bank to store energy. These batteries can be charged throughout the day and then used at night or on cloudy days to keep your essentials running.
- Wind Power

    - If you live in a windy urban area, a small wind turbine can complement your solar setup. Wind turbines are especially useful in places where solar energy might be limited during the winter months. Wind and solar together can create a more consistent power source.

- Backup Generators
    - For those times when you don't have access to renewable resources (or when the weather's just not cooperating), having a backup generator that runs on propane or natural gas is a good idea. Just make sure to keep fuel reserves on hand.

Off-Grid Water Systems: Collecting, Filtering, and Storing

When municipal water systems fail, clean water becomes one of the most urgent needs. Luckily, there are several ways to keep water flowing in your urban home:

- Rainwater Harvesting
    - Set up a rainwater collection system on your roof, where you can channel water into storage tanks. Once collected, you can filter and purify it for drinking, bathing, or cooking. If you live in an apartment or have limited roof access, you can still set up rain barrels at ground level to catch water runoff from gutters.
- Water Filtration Systems
    - If you're relying on collected rainwater or even local rivers or lakes, you'll need a solid filtration system. Consider reverse osmosis filters, UV purifiers, or simple carbon filters to remove contaminants.
- Water Storage
    - Having a large supply of stored water can be a lifesaver. Use food-grade water containers to store water in bulk. If you're living in an urban environment where space is limited, store water in large 5-gallon jugs or smaller bottles and keep them in a cool, dark place.

In my experience, during the flooding in my neighborhood, my rainwater collection system allowed us to keep basic water needs covered while municipal systems were down. It wasn't drinking water, but it worked well for cleaning and flushing toilets, which kept the home running smoothly until things cleared up.

Off-Grid Food Systems: Growing, Foraging, and Preserving

When grocery stores close and food supply chains break, growing your own food becomes a game-changer. In an urban environment, you can make this work with a little ingenuity.

- Urban Gardening
    - Even if you have a small balcony or windowsill, you can grow herbs, leafy greens, and even small vegetables. Start with containers, vertical gardens, or hydroponic systems to maximize your space.
- Community Gardens
    - If you have more space in your neighborhood, set up a community garden. This can be a great way to pool resources, share land, and grow food collectively. Be sure to compost the food scraps to improve soil quality and reduce waste.
- Foraging
    - Don't forget about what's growing wild around you. In many urban areas, you can forage for things like wild herbs, mushrooms, and fruit trees. Keep an eye out for these hidden resources and preserve them when you can.
- Food Preservation
    - Once you've harvested food, you need a way to preserve it. Get familiar with canning, freezing, and drying methods. These preservation techniques will allow you to store food for months and keep your pantry stocked even when resources are scarce.

Project: Build an Urban Survival Hub

Now it's time for a hands-on project to help you implement some of these ideas. The goal here is to build an urban survival hub that's capable of withstanding power, water, and food disruptions.

Identify a space that can act as your survival hub. This could be a community center, a local park, or even a neighborhood block. The goal is to have a central location where you can gather, share resources, and help each other in times of need.

Keep the following in mind:

- Install solar panels or set up a battery bank to store energy.

- Set up rainwater collection systems, and make sure you have filtration methods ready.
- Start a community garden and set up basic preservation systems (canning, drying, etc.).
- Set up community radios, or create an alternative communication system (like a phone tree or written notices) to keep everyone informed.

Key Points To Remember:

1. After a disaster, assess the damage to your home, starting with the foundation, roof, and electrical systems.
2. Scavenge for local materials such as wood, stone, and metal for repairs when traditional building supplies are unavailable.
3. Off-grid power systems like solar panels, wind turbines, and backup generators are essential for maintaining energy in a post-disaster environment.
4. Off-grid water systems such as rainwater harvesting, filtration systems, and water storage are key to ensuring access to clean water when municipal systems fail.
5. For long-term food security, set up urban gardens, community gardens, and foraging systems.

BOOK 30: BUILDING AN EMERGENCY FUND FOR SURVIVAL

Chapter 1:
Importance of an Emergency Fund

When life throws a curveball, the last thing you want to worry about is running out of money. Having an emergency fund is one of the most critical components of your survival strategy, and it's just as essential as having food, water, and shelter in my opinion,

How an Emergency Fund Supports Long-Term Survival

An emergency fund doesn't just help you weather the storm in the short-term. In fact, it's one of the cornerstones of long-term survival, especially in situations where access to funds or resources becomes limited.

In times of widespread crisis supply chains can break down. Grocery stores run out of food. Gas stations stop offering fuel. Stores close, and people are left scrambling for essentials. Without a stash of emergency cash, you might find yourself unable to buy what you need when it's available.

Having an emergency fund allows you to buy supplies in bulk when needed, like non-perishable food, water, or medications, before they run out. This proactive approach can mean the difference between staying well-stocked and finding yourself in the middle of a crisis without the resources you need to survive.

During the 2020 lockdown, stores were stripped bare in the first few days, and some people were left with nothing because they didn't have the foresight to plan ahead. I remember being able to walk into a store with cash and buy exactly what I needed, while others were left scrambling when stores sold out of essentials.

Building Your Emergency Fund

Building an emergency fund doesn't have to be overwhelming. It's all about starting small and making steady progress.

Here are some steps I suggest to get you started:

Step 1: Set a Goal

A common goal for an emergency fund is to have three to six months' worth of living expenses saved up. This gives you enough cushion to handle short-term disruptions. If you're just starting out, aim for $500 to $1,000 as a starting point and build from there.

Step 2: Start Small and Automate

Start by setting aside a small amount each month, whether it's $50, $100, or whatever fits your budget. Automate transfers into a separate savings account so you don't have to think about it. Over time, these small contributions will add up.

Step 3: Cut Unnecessary Spending

Review your spending habits. Are there places where you can cut back? Maybe it's skipping the daily coffee run or canceling a subscription you rarely use. Every little bit counts when you're working toward building a financial cushion.

Step 4: Prioritize Your Fund

When things get tight, it can be tempting to skip contributions to your emergency fund in favor of other expenses. But building that fund should be a priority. Treat it like any other bill—pay yourself first and then handle the rest.

Chapter 2:
Building Your Fund with Low-Risk Strategies

Now that you know why having an emergency fund is so crucial for survival and long-term security, let's get to the nitty-gritty of how to actually build it. The key to growing your savings is patience and consistency.

Budgeting for Emergency Savings

You might be thinking, "Where do I even start?" The good news is that building an emergency fund doesn't require fancy financial maneuvers or high-risk investments. In fact, budgeting is your best friend when it comes to setting money aside for unexpected events.

Before you start saving, you need to get an idea of where your money is going. Take a month and track every penny you spend. This will give you a clear picture of your income vs. expenses, and you'll see exactly where you might be able to trim the fat.

- Start by identifying your fixed expenses, like rent/mortgage, utilities, groceries, and transportation. These are the things you absolutely need to cover each month.
- Now, look at your discretionary spending—dining out, entertainment, subscriptions, shopping sprees. These are areas where you can cut back temporarily to build your emergency fund.

I learned this lesson the hard way in my youth when I was living paycheck to paycheck after a job change. I didn't realize how much I was spending on impulse buys and subscriptions I barely used. After tracking my expenses, I realized I could save a couple hundred dollars every month just by cutting out the extras. It was a game-changer for me when I redirected that money to my emergency fund.

Set Realistic Savings Goals

Once you've identified your expenses, it's time to set some realistic savings goals. Don't try to save an unrealistic amount too quickly. Build your emergency fund in manageable steps. A good goal for most people is to save three to six months' worth of expenses.

Start small! Aim for saving $500 to $1,000 to get started, and then work your way up to your full goal. Don't forget: your emergency fund is there to give you peace of mind, not stress you out with unreachable targets.

The easiest way to make sure your savings goal is met is by automating your transfers. Set up an automatic monthly deposit from your checking account to your emergency fund savings account. Even if it's just $25 or $50 a month, that consistency adds up over time.

Using High-Interest Accounts for Growth

Now that you've set a budget and automated your savings, let's talk about how to make your emergency fund work for you while it's sitting there. You don't want it to just sit idle in a regular checking account. You want to maximize its growth while keeping your money safe. High-interest savings accounts and certificates of deposit (CDs) are great tools to help your money grow with minimal risk.

High-Interest Savings Accounts

A high-interest savings account is a perfect place to park your emergency fund because it keeps your money safe and easily accessible, but it also earns interest over time. The goal is to find an account with a higher interest rate than the average savings account, so your money grows while you aren't using it.

- Online Banks
    - Many online banks offer better interest rates than traditional brick-and-mortar banks. Check out options that provide anywhere from 1% to 2% APY (annual percentage yield). It might not sound like much, but the power of compound interest means that over time, your savings will grow.
- Access
    - High-interest savings accounts give you easy access to your money when you need it, which is essential for emergency funds. If you're looking for something with higher

growth potential but less accessibility, you could look into Certificates of Deposit (CDs), which offer higher rates but require you to lock up your funds for a set period.

Chapter 3:
Protecting and Growing Your Emergency Fund

So, you've built your emergency fund, and now it's time to talk about the next steps like protecting and growing that money. While having a fund in place is already a huge accomplishment, you don't want to just let it sit around collecting dust. The goal here is to maximize the value of your emergency fund while also ensuring that it remains safe and accessible when you need it the most.

Diversifying Savings and Investments

Once you've got a solid emergency fund in place, the next question is: "How can I make that fund work for me?"

Diversifying your savings and investments is an important part of growing your wealth while still keeping a safety net for unexpected crises. The goal here is to balance liquidity (the ability to access your money quickly) with growth potential.

The principle behind diversification is simple: Don't put all your eggs in one basket. If you invest in only one asset (like a savings account or one stock), you're at risk if that asset loses value or fails to perform. By spreading your money across different investment vehicles, you help protect your emergency fund while also giving it the chance to grow over time.

If you're looking to grow your fund over a longer period and you don't need access to the money immediately, consider low-risk investments that can earn higher returns. These are best for funds you can afford to leave untouched for a while, beyond your emergency savings.

- Index Funds or ETFs (Exchange-Traded Funds)
    - These are low-cost investments that track the performance of the overall market. They offer good growth potential over time and are less risky than picking individual stocks.
- Bonds

  - Bonds are a good way to earn steady returns without much risk. You lend money to a corporation or government, and in return, you get interest payments. They can be a stable addition to your investment portfolio.

Project: Create a Monthly Financial Tracking System

A critical step to ensuring your emergency fund grows, and that you are financially prepared for unexpected events, is setting up a system to track your finances regularly. Here's how to build a monthly financial tracking system that helps you stay organized and on top of your savings goals.

Step 1: Track Your Income and Expenses

Start by tracking every dollar coming in and going out. Use an app like Mint, YNAB (You Need a Budget), or even a simple spreadsheet to log your income and expenses. This will give you a clear idea of where you're overspending and where you can cut back.

Step 2: Set Monthly Budget Categories

Categorize your spending:

- Essentials: Rent/mortgage, utilities, groceries, transportation
- Savings: Emergency fund, retirement savings
- Discretionary: Entertainment, dining out, shopping

Setting a budget will help you control your spending and ensure you're putting money into your emergency fund and other financial goals every month.

Step 3: Review and Adjust Regularly

At the end of each month, review your finances. Did you meet your savings goals? Where can you improve? Look at your discretionary spending. Could you cut back on dining out or limit subscriptions? Adjust your spending habits as needed to meet your financial targets.

Key Points to Remember:

- An emergency fund is crucial for navigating life's curveballs, giving you the financial cushion to get through tough times like supply shortages or unexpected expenses.
- Start small with a savings goal of $500 to $1,000, automate your deposits, and treat your emergency fund like a bill you can't skip.

- Track your spending, cut back where you can, and consistently stash away money to build your emergency fund—little by little, it adds up.
- Use high-interest savings accounts or CDs to let your emergency fund grow over time while still keeping it accessible for when you need it most.
- Diversify your savings with low-risk investments like index funds or bonds to help your emergency fund grow without risking it all.

CONCLUSION

Congratulations!

You've made it through this journey of learning how to prepare and thrive when the unexpected happens.

This book was meant to give you a foundation for building a life that's resilient, resourceful, and ultimately ready for any unexpected crises that comes your way.

But now, the real work begins. The knowledge you've gained here isn't the end of your preparedness journey. It's the beginning! The key to truly thriving in any crisis is to take what you've learned and apply it (bit-by-bit) into your everyday life.

Keep in mind that you don't need to implement everything all at once. Start small. Each step you take will move you closer to the life of self-sufficiency and security that you're aiming for. It's not about doing it perfectly. It's about starting where you are and doing what you can.

Keep in mind that preparation isn't a one-time event. It's an ongoing process. You'll need to adapt as the world around you changes and as you learn more. The information in this book is just a snapshot of the many strategies you can use to safeguard yourself and your loved ones.

There's always more to discover, so continue your research, keep seeking new information, and learn from others in the prepper and survival communities. The more you know, the better prepared you'll be when the time comes to rely on your skills.

Made in United States
Orlando, FL
04 November 2026

90202585R00175